Introduction
to
Protein Sequence Analysis

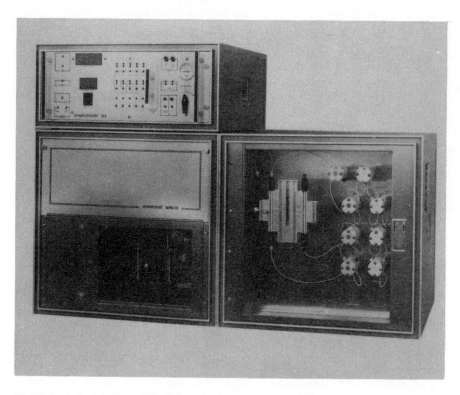

The Sequemat Mini-15 Solid Phase Sequencer. This instrument is able to determine a minimum of fifteen residues per day of a suitable derivatized peptide or protein. It possesses two columns so permitting the simultaneous sequence analysis of two samples. (Reproduced by courtesy of Sequemat Inc., Watertown, Massachusetts, USA.)

Introduction
to
Protein Sequence
Analysis

L. R CROFT
Lecturer in Biochemistry
University of Salford

JOHN WILEY & SONS
Chichester · New York · Brisbane · Toronto

British Library Cataloguing in Publication Data

Croft, Laurence Raymond
  Introduction to protein sequence analysis.
  1. Proteins — Analysis
  2. Amino acids
  I. Title
  547'.75   QD431   79-41488

  ISBN 0 471 27710 X

Typeset by Preface Ltd, Salisbury, Wilts. and printed by Page Bros. (Norwich) Ltd., Norwich

To
Ian

Contents

Preface

I remember showing an interest in the field of amino acid sequence determination soon after gaining my Ph.D. in 1968 and being advised by many well-intentioned biochemists to avoid this area since it was considered that the advent of the sequencer had made it virtually a closed book. How wrong this advice was for this field has gained a position of great importance in present biochemistry, well beyond the expectations of little over a decade ago. However, as must be stated at the outset, sequence determination is not everyone's cup-of-tea: to many biochemists it is a necessary evil that must be done as quickly and as painlessly as possible. I have much sympathy with their attitude. There is, however, a small minority of biochemists that thrive in the systematic and ordered world of amino acid residues and peptide fragments. (No doubt this may one day be the subject of a sociological study.) The fact is, as may be clearly seen from the size of the companion volume *Handbook of Protein Sequences*, the workers in this field have made a significant and long-lasting contribution to biochemistry. The recent magnificent achievement of the sequence of β-galactosidase, consisting of 1021 residues, must be a pinnacle of endeavour for years to come. Yet the question has been raised that this is the beginning of the end for protein sequencing. This is a matter of opinion. My own feeling is that the complementary nucleic acid sequencing field will never replace amino acid sequencing, which will remain for years to come as important as ever. It will be possible a decade hence to say, as W. A. Schroeder, in 1968, 'the number of proteins for which a full sequence can be written with reasonable assurance is a mere handful compared to those that have names—let alone those that exist'.

It was during my many discussions with students studying both chemistry and biochemistry that I became aware that the approach taken by many textbooks and lecturers to this field of biochemistry was out of date, and the few textbooks that were available to specifically cover this field had been published several years ago were also inadequate. It is therefore my hope that this small book will bridge the gap and will provide the student and research worker with an up-to-date view of protein sequencing.

Further, I hope that this book will inspire a further generation of researchers to enter this field which still offers much unexplored territory. I hope in particular that this guide will find a home in many of the colleges and laboratories of third-world countries and encourage workers in these places to enter this field.

Throughout the text I have not hesitated to express my own opinions and to draw on my own experiences of sequencing, which has in the main been with lens proteins. I have at several times in the text expressed my view that sequencing although possible with expensive and sophisticated equipment is nevertheless still practicable using very simple and cheap facilities. This is not meant in anyway to denigrate the technological achievements of the particular manufacturers of these instruments but simply my attempt to encourage the research worker who has not the benefit of such equipment. I hope the particular manufacturers concerned, who have given me every assistance in writing this book, will understand my approach.

Salford   L. R. CROFT

September 1979

Abbreviations

Ala   Alanine
Arg   Arginine
Asn   Asparagine
Asp   Aspartic acid
Asx   Aspartic acid or asparagine (undefined)
Cys   Cysteine
Glu   Glutamic acid
Gln   Glutamine
Glx   Glutamic acid or glutamine (undefined)
Gly   Glycine
His   Histidine
Ile   Isoleucine
Leu   Leucine
Lys   Lysine
Met   Methionine
Phe   Phenylalanine
Pro   Proline
Ser   Serine
Thr   Threonine
Trp   Tryptophan
Tyr   Tyrosine
Val   Valine
CMC   S-carboxymethylcysteine

Introduction

Knowledge of the complete amino acid sequence of a protein may be desirable for one or more reasons. If the protein is an enzyme, it may be in order to obtain an understanding of its mechanism of action, or it may be a means for studying the evolutionary history of life. It is now well established that proteins are useful tools with which to trace the thread of evolution and they are in a sense 'ultramicrofossils'. Such studies are now vigorously pursued on a world-wide scale (Table 1).

Further, the protein might be of immediate medical interest, and the sequence needed to form the basis for a synthetic programme, or an investigation as to its origin. The recent enkephalin 'bandwagon' is a good example of this. Agricultural archaeology is another area in which the importance of amino acid sequence information is now well established, and an understanding of early selective breeding has been derived from analytical studies of ancient protein material, particularly the collagens of bone, tendon, or skin, and the keratins of hair, which may be perfectly preserved through 20 000 years, and indeed much longer.

Table 1. The origin of sequence information published during 1977 and 1978

Country where the work was performed	% of the total
U.S.A.	29
Japan	13
W. Germany	12
South Africa	10
U.K.	8
France	6
Holland	6
Belgium	3
Australia	3
Sweden	3
U.S.S.R.	2
Italy	1
Taiwan	1
Canada	1
Switzerland	1

2

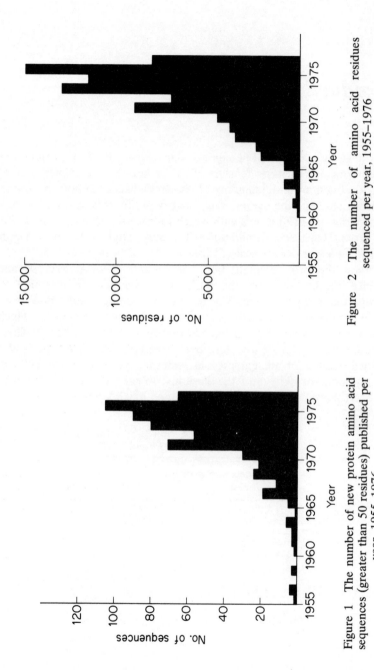

Figure 1 The number of new protein amino acid sequences (greater than 50 residues) published per year, 1955–1976

Figure 2 The number of amino acid residues sequenced per year, 1955–1976

It is only twenty-five years ago that the determination of an amino acid sequence of a protein was looked upon as an impossible task, yet during this period the methods of sequence determination have been developed to such a sophistication that today this task, to many, is considered routine. If we look at the number of sequences deduced per year over the last 25 years we can clearly see that this endeavour up to about 1973, was at an explosive stage of development, and that after this time the number of sequences deduced levelled off to an approximate doubling of information every two or three years (Figures 1 and 2). Yet as a whole the development of protein structure determination has been painfully slow if compared to other fields of scientific endeavour.

HISTORICAL PERSPECTIVES

It was about the turn of the century that the 'peptide hypothesis' of protein structure was put forward by Hofmeister and Fischer. Emil Fischer applied the methods of organic chemistry to the problem of protein structure, and obtained from the hydrolysates of various proteins many peptides in crystalline form, which were compared with synthetic samples. These early triumphs firmly established the basic concepts of protein structure, but in the years that followed, it was clear that the methods of classical organic chemistry were inadequate, and fifty years were to elapse before any further progress was made. By this time the techniques of paper chromatography and electrophoresis had been introduced as a means for checking homogeneity and confirming identity, and it was by using these methods that Sanger was able to determine the complete amino acid sequence of insulin in 1955. This was the culmination of a decade of work by Sanger; the choice of insulin, a small protein of 51 residues and readily available in large amounts, was indeed fortunate. If a larger protein had been chosen for sequence analysis, for example haemoglobin, the work could very well still be in progress, if not abandoned many years ago. Sanger's achievement must surely rank as one of the great landmarks in science this century, as it provided the proof that proteins indeed had unique structures and could be regarded as molecules and not vague colloids. The development of the genetic code was a direct consequence of Sanger's work and subsequently led to the advent of the new science of molecular biology. It can now be clearly seen that what Sanger did was to be courageous enough to apply new, and at the time controversial, methods to this particular area of natural product chemistry. Chromatographic and electrophoretic criteria were employed for confirmation of purity in place of the usual crystallization and melting point tests. It was subsequently shown by chemical synthesis, and X-ray crystallography, that this approach was valid, and comparable to other more classical methods of structure determination, and the way was then open for more people to enter and develop this field. And how the field has developed is illustrated in Tables 2 and 3; the first shows how techniques

Table 2. Amount and estimated time taken to sequence a hypothetical peptide having 50 amino acid residues, at four different times during the last 30 years

Year	Amount required	Time taken
1950	100 g	10 years
1960	10 g	3 years
1970	1 g	1 year
1980	1 mg	1 week

have improved in sensitivity and how the time taken to sequence a hypothetical 50-residue peptide has decreased during the last three decades.

The second (Table 3) lists some of the major landmarks in this field during the last three decades. Progress was enhanced by the introduction of two important developments, namely, the automatic amino acid analyser, which replaced the rather subjective visual quantification of amino acids that had previously been derived from paper chromatograms, and the ingenious stepwise degradation procedure devised by P. Edman. Both of these were catalysts for the rapid growth of this area of biochemistry. It is interesting to compare development in the United States with that of the United Kingdom. Although the initial work on sequencing took place in Britain, it can be clearly seen that the lead was quickly taken over by American workers who were more willing, and financially able to adopt automated procedures, such as the amino acid analyser. It is still true that in Britain it is much easier to raise funds for personnel than it is for the purchase of sophisticated equipment, and this is only too evident when one compares the number of automatic sequencers in university biochemistry departments in

Table 3. Landmarks in protein sequence determination in the last 25 years

Year	Protein Sequence	Number of residues
1955	Insulin	51
1961	Haemoglobin beta chain	145
1962	Haemoglobin alpha chain	141
1962	Cytochrome c	105
1963	Ribonuclease	124
1963	Lysozyme	129
1965	Myoglobin	153
1965	Virus coat protein	159
1966	Trypsinogen	216
1967	Glyceraldehyde 3-phosphate dehydrogenase	340
1969	Immunoglobin γ-chain	446
1973	Immunoglobulin μ-chain	537
1975	Serum albumin	585
1978	β-galactosidase	1021

the U.K., to that in the United States. However, there have been advantages with this situation, as scientists are much more ready to innovate when a particular piece of machinery is unavailable. A particular instance is the development of the dansyl-Edman procedure of Gray and Hartley. This without question is a major rival to the automatic sequencer, yet is so facile.

BASIC APPROACH TO PROTEIN SEQUENCING

Before work on the amino acid sequence can even be considered the protein under investigation must be established to be homogeneous. Discussion on the methods available for the purification of proteins is considered to be outside the scope of this small book; there are however many excellent reference books available to adequately cover this field, which has during the last decade altered very little. Some discussion, however, will be given concerning what criteria should be used to assess the purity of a protein preparation.

It is true to say that the only real proof that a protein is homogeneous is that it possesses a unique amino acid sequence, so at the commencement of a sequence investigation it is impossible to know that the protein is 100% pure. At best, therefore, one can only apply a number of tests, none in themselves being infallible, but when taken together, some confidence as to the protein's purity should be possible. The following tests should be included:

(1) The protein should be homogeneous in size. Molecular weight estimates may be obtained from ultracentrifuge studies, gel filtration measurements, or by DISC-gel electrophoresis in the presence of sodium dodecyl sulphate.

(2) The protein should be homogeneous in charge. This information may be readily derived from electrophoretic studies using a variety of media (cellogel, polyacrylamide gels, startch gels and agar gel) and over a pH range. The more sensitive technique of isoelectric focusing should also be employed.

(3) The protein should have a unique amino acid composition, that should not vary during any further attempt at purification.

(4) The protein should have unique amino- and carboxyl-terminal residues. Quantitative measurements of these residues should give an estimate of the molecular weight, which should agree with that previously determined.

(5) Any further attempt to purify the protein should not lead to an increase in its biological activity. Any decrease observed would probably indicate partial denaturation during the purification step.

(6) It should be possible to crystallize the protein.

During the author's investigations on lens proteins, one particular protein, γ_{IV}-crystallin, was found to fulfil all the above criteria, yet during subsequent sequence investigations it was found to consist of at least three homologous proteins. Such are the frustrations of sequence projects and it is well to begin any such work by a preliminary investigation as to the sequences around particular regions of the molecule, such as for example the cysteine residues. The results from such a study would give confidence to proceed further with the investigation.

Having taken reasonable steps to ensure that the protein under investigation is reasonably pure, it is then possible to embark on the actual sequence determination. This will entail the following stages:

(1) The amino acid composition of the protein is determined and an investigation as to whether there are any other substances covalently attached to the protein, such as carbohydrate, is carried out.

(2) The amino acid sequences at the amino- and carboxyl-termini are determined.

(3) The protein is treated with a specific endoprotease, or chemical reagent, to fragment it into a number of peptides. The actual one used will depend on the overall strategy. Two approaches are possible: first a large number of small fragments might result, with the consequent problems of separation and purification; secondly, a small number of large fragments might result, which usually present solubility problems, as large hydrophobic peptides are usually insoluble in aqueous media, unless one resorts to the messy and expensive use of 6 M urea solutions. Probably the most useful cleavage procedure is tryptic digestion.

(4) Each of the peptides produced from (3) are purified and their amino acid compositions determined. These should add up to the overall amino acid composition of the protein. Failure to do so would probably indicate missing peptides.

(5) The amino acid sequence of each of the purified peptides is determined. A comparison of the methods available is given in Table 4.

Table 4. Comparison of conventional methods for amino acid sequence determination

Method	Size of suitable peptide (residues)	Approximate amount needed (nmol)	Comments
Manual methods	10	50	Simple and cheap
Automatic:			
(a) soild phase	10–15	100 ⎫	requires costly
(b) liquid phase	20	250 ⎬	equipment, and
Mass spectroscopy	5–10	100 ⎭	specialized expertise

(6) Stages (3)–(5) are repeated using a reagent of different specificity in order to overlap the initial peptides. This might well be cyanogen bromide.

(7) It should be possible to assemble the complete sequence from the information gained in these experiments, however further reagents may be employed to elucidate the sequence at problematical regions.

In recent years the strategy has somewhat changed, with more emphasis on the use of automatic procedures, which work best on large peptide fragments. Thus the fragmentation of a protein into a small number of large peptides is normally the first step. For example, a protein of 160 amino acid residues could be fragmented into four peptides each of approximately 40 residues. Each of these could be completely sequenced in the sequencer, and when this information is combined with knowledge of the sequences of the terminal regions a complete sequence should be realized.

FUTURE PROSPECTS

Microsequencing

These developments have occurred to enable biochemists to investigate minute quantities of biologically active proteins, such as histocompatibility factors and prohormones. Several approaches have been proposed but they have in common the use of radioactively labelled protein to increase the sensitivity of established procedures. The protein is labelled either by chemical or biological means. In the case of biological incorporation of radioisotope, this may be achieved by allowing translation of the protein mRNA to occur in a reticulocyte lysate containing a mixture of radioactive amino acids. Alternatively radioactive amino acids may be incorporated by incubation of living cells synthesizing the protein of interest; however, the problem here is the difficulty of getting the non-essential amino acids labelled. The chemical approach has been to use ^{35}S-phenylisothiocyanate of high specific activity in a preliminary coupling step in the Edman degradation. The released thiazolines after conversion to their respective PTH-amino acids are mixed with unlabelled carrier PTH-amino acids, and separated by TLC and identified by autoradiography.

Microsequencing therefore promises to be generally applicable to the determination of the primary structure of trace quantities of biologically important proteins, and has in the order of 10 000 times the sensitivity of conventional methods.

Nucleic Acid Sequencing

Techniques for the sequence determination of nucleic acids have increased beyond all expectations during the last decade, and so too have the techniques for the isolation of specific mRNA molecules. Considering the relative ease by which the sequences of nucleotides may be determined in these products, it has been looked upon by some workers as an alternative procedure whereby protein sequence determination might be arrived at. This, however, is somewhat a controversial area at the present time of writing, and it does not seem feasible that this approach will replace the classical methods of protein sequencing within the near future.

ABBREVIATIONS AND CONVENTIONS

As information on the amino acid sequences of proteins has increased, methods have been devised to abbreviate and simplify this information. To begin with the term 'amino acid sequence' is synonymous with the 'primary structure' of a protein, these terms being completely interchangeable. However, the covalent structure of a protein refers to the location of all the covalent bonds present in a protein molecule, and may include location of disulphide bridges, and the position of prosthetic groups, and thus has not the same meaning as the former two terms.

The primary structure of a peptide is normally presented as a sequence of amino acid residues using the well-known three-letter code to represent these (see Abbreviations).
For example,

Ala–Gly–Ser

represents a tripeptide formed between the amino acids alanine, glycine and serine. It may also be written as

Ala.Gly.Ser

or simply,

Ala Gly Ser (also ALA GLY SER).

By convention the residue on the far left, namely alanine, is always the N-terminal residue, i.e. it possesses a free α-amino group, and the residue on the far right is the C-terminal residue, i.e. has a free α-carboxyl group, in this instance the amino acid serine. The peptide bond is represented by the hyphen, or period, or simply a space, whereas a comma indicates that the sequence of the amino acids is not known.

CHAPTER 1

Enzymic Cleavage of Proteins

The early studies on the amino acid sequences of proteins were performed using partial acid hydrolysis as a method of degradation, for there was a reluctance amongst biochemists to employ proteolytic enzymes for fear of transpeptidation. These misapprehensions were soon realized to be without substance, and the use of enzymes in this field became widely accepted. Nevertheless, in more recent times, as a consequence of the trend toward automatic sequencing and the need for large peptide fragments, biochemists have looked again at proteolytic enzymes to find ways of limiting their activity, or to seek new, more specific ones. Chemical methods, at one time quite ignored by the majority of workers, are once again in fashion. Thus at the present time this field is in a state of flux, and for the purpose of this section only the proteases that in the past have been most useful will be considered, together with a summary of the recently introduced enzymes.

GENERAL PROCEDURES FOR ENZYME HYDROLYSIS

In all enzyme reactions there are four main factors to consider, namely

(1) pH,
(2) temperature,
(3) the enzyme: substrate ratio, and
(4) the period of incubation.

The optimum conditions for a particular hydrolysis are best found by trial and error, following the progress of the reaction by peptide mapping.

Preparation of the Protein for Enzyme Digestion

Native proteins are not generally susceptible to enzymic hydrolysis. There are cases, however, where a susceptible bond might be accessible to the enzyme active site and be cleaved, possibly leading to several large fragments; such specific reactions are unusual, for what normally happens in these circumstances is a slow denaturation and hydrolysis of the protein substrate—which is unsatisfactory. The protein must therefore be denatured prior to enzymic digestion; this process, at one time a complete mystery to biochemists, is now thought to involve the unfolding and destruction of the

9

secondary and tertiary structures of the macromolecule, converting all conformations present to a random coil rigid structure. Such a configuration is readily amiable to enzymic hydrolysis.

Methods of Protein Denaturation

(i) Heating: boiling in a water bath for between 10–60 minutes; the protein usually precipitates under these conditions and is collected by centrifugation.

(ii) Treatment with strong acid: (e.g. trichloroacetic acid) the protein is precipitated from aqueous solution with 10% trichloroacetic acid, collected by centrifugation and washed several times with water.

(iii) Performic acid oxidation: the protein is treated with an excess of HCOOOH at 0°C for about 2 hours, after which it is precipitated from

Table I. Useful proteolytic enzymes

Enzyme	Source	Specificity: cleavage on the		Optimal pH
		C-terminal side of	N-terminal side of	
Trypsin	Ox pancreas	Arg, Lys (except when Pro next residue)		8–9
Chymotrypsin	Ox pancreas	Trp, Tyr, Phe, Leu and to a lesser extent Met, Asn, Gln, His (except when Pro is the next residue)		8–9
Thermolysin	*Bacillus thermoproteolyticus*		All hydro- phobic amino acids except e.g. -*x*-Leu-Pro- not cleaved	8
Pepsin	Ox pancreas	Similar but less specific than chymotrypsin, cleaving mainly on the N- and C- sides of aromatic amino acids and leucine		2

solution by 25% trichloroacetic acid, collected by centrifugation and washed thoroughly. This treatment oxidizes the cystine/cysteine residues to cysteic acids, the methionine to methionine sulphone, and the tryptophan is completely destroyed.

(iv) S-carboxymethylation: the protein in 6M-urea is treated with a reducing agent to convert the cystines to cysteine residues, and these are modified by either iodoacetic acid or iodoacetamide, at pH 6–8, in the dark, for about 1 hour. The protein is obtained after dialysis by lyophilization.

(1) TRYPSIN

Specificity

Trypsin has been the most useful of all the proteolytic enzymes, because of its high specificity, cleaving only on the C-terminal side of lysine and arginine residues. It has been used to study the number and nature of the subunits of many proteins, using the simple technique of peptide mapping, the number of peptides being equal to the sum of the lysine and arginine residues plus one (the C-terminal peptide), in the case of identical subunits. However, it should be borne in mind that repetitive sequences of either lysine or arginine are only cleaved slowly, resulting in an extra peptide, it is thought that this is due to the influence of the polar side chain on the cleavage of the bond to be cleaved. Kasper (1975) reports that the sequence Arg–CySO$_3$H is only partially cleaved; however, the present author has found the same sequence to be rapidly cleaved by trypsin, in the lens protein γ-crystallin, and it is clear that the local environment affects the susceptibility of a particular bond. Bonds involving the amino acid proline are not cleaved, but there are exceptions

'Contaminating' Chymotryptic Activity

Some commercial preparations of trypsin may contain small amounts of chymotryptic activity, which, during the course of a digestion, may produce unwanted peptides, as well as reducing the yield of the tryptic peptides. Thus, cleavage of the bond between Tyr[28]–Phe[29] in γ-crystallin was always observed (Croft 1972a, b), and a similar cleavage was found by Enfield *et al.* (1975) in their study of rabbit muscle parvalbumin: in this case the bond Phe[18]–Ala[19] was always cleaved. Chymotryptic activity can be reduced by treating the trypsin preparation with a chymotrypsin inhibitor such as L-(1-tosyl amido-2-phenyl) ethyl chloromethyl ketone (TPCK); however, Richardson *et al.* (1978), in their work on the sequence of the α-subunit of pea lectin, always found the Tyr[4]–Thr[5] bond to be cleaved, even after the trypsin had been treated with TPCK, and they had also, as an added precaution, treated the trypsin with 1 mM-HCl to further inactivate any residual chymotrypsin activity. It is interesting that a similar anomalous cleavage has been reported to occur during the tryptic digestion of lentil lectin, a protein homologous to the pea protein (Foriers *et al* 1978).

Aspecific Cleavage

The reports mentioned above clearly indicate the important role the substrate has on the course of the enzyme digestion, and this idea has been further substantiated by the finding of van Den Berg *et al.* (1977) that many aspecific cleavages were found in tryptic digests of guinea pig ribonuclease A, that were not found with guinea pig ribonuclease B, even though the same preparation of trypsin was used. One unusual cleavage was between Lys^{113}–Pro^{114}, and this was found also in digests of ox seminal ribonuclease. That trypsin can, in certain circumstances, hydrolyse the bond between Lys(or Arg) and Pro, is now well established. Bennett *et al.* (1978), in their study of dihydrofolate reductase, found that the bond between 52–53 was cleaved by trypsin: the sequence around the susceptible bond was –Gly^{51}–Arg^{52}–Pro^{53}–Leu^{54}–, and a similar hydrolysis was found by Croft (1972a, b), between residues 48 and 49 in the sequence –Gln^{47}–Arg^{48}–Pro^{49}–Asp^{50}–, of the lens protein γ-crystallin. Carnegie (1969) also found that trypsin was able to cleave the Arg–Pro bond, in the sequence –Gln–Arg–Pro–Gly–, during his investigations on the encephalitogenic protein from human myelin. This protein also contained the interesting methylated derivatives of arginine, namely ω-N-monomethyl Arg, and ω-N,N'-dimethyl Arg, which resisted tryptic hydrolysis (Dunkley and Carnegie 1974); however, some unusual amino acids in proteins do render the peptide bonds susceptible to cleavage by trypsin, for example S-aminoethyl cysteinyl bonds are cleaved by trypsin, though not as rapidly as its lysine analogue.

Limitation of Cleavage by Chemical Modification

As trypsin can cleave on the C-terminal sides of both lysine and arginine, it is clear that its usefulness could be enhanced if its activity could be limited to one or other of these residues. Many chemical procedures have been introduced to this end. Most have been directed toward the ε-amino group of lysine (being simpler to modify than the guanidinium group of arginine), and the most successful has been citraconic anhydride, introduced by Dixon and Perham (1968), having the added bonus that it is readily removed. This reagent has been used to identify and isolate the N-terminal peptide from

(N–citraconyl)Gly–(ε–N–citraconyl) Lys-Ile-Thr-Phe-Tyr-Glu-Asp-Arg
(Peptide *N*)

pH 2.0
20 hours

Gly-Lys–Ile–Thr–Phe–Tyr–Glu–Asp–Arg
(Peptide *N'*)

Trypsin

Gly–Lys
(Peptide *N'a*)

Ile–Thr–Phe–Tyr–Glu–Asp–Arg
(Peptide *N'b*)

Figure I   Isolation of the N-terminal peptide from γ-crystallin
using citraconic anhydride to limit the cleavage with trypsin to
arginyl peptide bonds

γ-crystallin (Croft 1972a). The citraconylated protein was digested with
trypsin and the very acidic peptide (*N*) isolated by paper electrophoresis.
The blocking groups were readily removed by incubation in dilute acetic
acid overnight, and the resulting peptide (*N'*) further treated with trypsin to
give peptides *N'a*, and *N'b* (Figure I).

A related procedure has been introduced for arginine residues by Patthy
and Smith (1975). Arginyl residues were found to be specifically blocked by
reaction with cyclohexane-1,2-dione in borate buffer (pH 8–9) to form N^7,
N^8 (1,2-dihydroxycyclohex-1,2-ylene) derivatives. These can be stabilized by

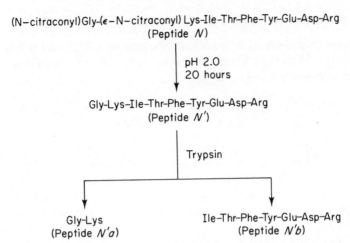

the presence of borate ions, the product being resistant to cleavage by
trypsin; however, the arginine is readily regenerated by mild treatment with
hydroxylamine.

General Reaction Conditions

The peptide or protein (200 nmol) is dissolved in 0.2 ml of 1% NH_4HCO_3. The enzyme (2 nmol, 0.05 mg) dissolved in 1 mM-HCl, is added and the mixture incubated for 3–4 hours at 37°C.

(2) CHYMOTRYPSIN

Chymotrypsinogen is the inactive precursor of chymotrypsin, and is a single-chain polypeptide of 245 amino acid residues, stabilized by five disulphide bridges. The zymogen has no proteolytic activity and is activated by trypsin, which hydrolyses the peptide bond between Arg^{15}–Ile^{16}, forming π-chymotrypsin, which acts upon itself to form the fully active α-chymotrypsin, liberating two peptides Ser^{14}–Arg^{15}, and Thr^{147}–Asn^{148} in the process. α-Chymotrypsin consists therefore of three chains A, B and C, containing 13, 130, and 96 amino acids respectively, joined together by two disulphide bridges. The specificity of this enzyme is not as high as that of trypsin, but nevertheless it has proved invaluable to the protein biochemist. In general the active enzyme cleaves preferentially the peptide bond on the C-terminal side of aromatic and bulky hydrophobic amino acids (Tyr, Trp, Phe, Met and Leu). However, cleavage may also occur on the C-terminal side of Gln, Asn, His and Thr. A high enzyme-to-substrate ratio may enhance these minor splits, and even bonds like Ala–Val are cleaved, as in Ferredoxin II from *Chlorobium limicola* (Tanaka *et al.* 1975). The rate of hydrolysis tends to be slowed down by the presence of polar groups—these may be terminal amino groups, or side-chain carboxylic acids—and it is wise to view the whole environment around a particular peptide bond when considering its susceptibility to this enzyme.

General Reaction Conditions

The peptide or protein (0.2 mol) is dissolved in 0.2 ml of 1% NH_4HCO_3, the enzyme (0.05 mg, 2 nmol) is added and the mixture incubated at 37°C for 3–4 hours.

(3) PEPSIN

The specificity of pepsin is similar to that of chymotrypsin but as its optimum pH is at pH 2, it has proved very valuable to the biochemist, particularly during investigations of the location of disulphide bridges, as in acid solution disulphide interchange reactions are minimal. It does, however, differ from chymotrypsin in its ability to cleave peptide bonds on both the N- and C-terminal sides of susceptible residues. In the main these are the aromatic amino acids and leucine; it does not hydrolyse bonds attached to proline residues.

General Reaction Conditions

The protein (100 nmol) is dissolved in 5% formic acid (0.1 ml) and to it is added pepsin (1μ1, of a solution prepared by adding 1 mg of pepsin to 1 ml of 0.5 M NaCl). The digestion is allowed to proceed at room temperature, for 1 hour.

(4) THERMOLYSIN

This is the least specific of the proteases so far discussed, and for that reason it has had only limited application in the fragmentation of proteins, the peptides produced being small and tedious to purify. It has, however, been of considerable use in the fragmentation of large peptides from a primary digest. Thermolysin is a metalloenzyme containing one Zn and three to four Ca atoms per molecule. The function of the Ca is to enhance the stability of the enzyme, and it is thus advisable to include a small amount of calcium ions in the reaction medium.

Thermolysin cleaves on the N-terminal side of hydrophobic amino acids, namely Ile, Leu, Val, Phe, Ala, Met and Tyr, however, if the carboxyl of the said amino acid is joined to the imino acid proline, hydrolysis does not occur. Needless to say, there are instances in the literature outside this category. Emmens *et al.* (1976) found the bond Thr[99]–Ser[100] in pike whale ribonuclease to be cleaved by thermolysin. The local environment plays an important role in these aspecific cleavages. Hydrolysis can occur on the N-terminal side of cysteic acid residues when the subsequent residue in the chain is lysine (Scheffer and Beintema 1974), and it is thought that the sulphonic acid side chain is shielded by the proximal ϵ-amino group of the lysine. And Matsubara *et al.* (1970) found thermolysin to cleave on the N-terminal side of glutamyl residues when it was adjacent to arginine.

General Reaction Conditions

The peptide (300 nmol) is dissolved in ammonium acetate buffer (0.5 ml, 0.2 M, pH 8.5, containing 5 mM $CaCl_2$) and the enzyme (0.1 mg) added in 100 μl of the same buffer. Reaction continues at 40°C for about 3 hours.

(5) RECENTLY INTRODUCED PROTEOLYTIC ENZYMES

There has been and still is a major search for highly specific proteolytic endopeptidases. Many have been described in the literature but few have been successfully used in amino acid sequence determination; Table II lists several new proteases that have been applied with some degree of success. These will not be considered here in detail, as they have not been used extensively enough to warrant a comprehensive evaluation, except the 'Glu' protease which in recent years has found widespread application. This

Table II. Recently introduced proteases

Enzyme	Source	Specificity: cleavage on the C-terminal side of	N-terminal side of	Reference
Clostripain	*Clostridium histolyticum*	Arg		Mitchell and Harrington (1968)
'Pro' protease	Lamb kidney	Pro		Fleer *et al.* (1978)
'Lys' protease	*Armillaria mellea,* Myxobacter		Lys	Gregory (1975); Jörnvall (1977)
'Glu' protease	*Staphylococcus aureus*	Glu		Houmard and Drapeau (1972)

enzyme cleaves on the carboxyl side of glutamyl residues and has proved very useful in the differentiation of Glu/Gln residues, and in providing the investigator with peptides possessing C-terminal glutamic acid, enabling facile attachment to resins for the solid-phase Edman degradation.

Staphylococcal Protease

This is one of the most interesting and useful of the many endopeptidases introduced in recent years. It was first isolated from *Staphylococcus aureus*, strain V_8, by Houmard and Drapeau (1972). These investigators found that this protease, with the remarkably low molecular weight of 12 000, could cleave on the C-terminal side of glutamyl and aspartyl residues when the hydrolysis is performed in phosphate buffer (pH 7.8). If, however, the buffer is changed to either NH_4HCO_3 (pH 7.8) or ammonium acetate (pH 4.0), cleavage only at glutamyl residues occurs. The nature of the amino acid attached to the carboxyl of the glutamyl residue, except proline, did not influence the rate of hydrolysis, and as glutaminyl residues were not cleaved at all it was concluded that the γ-carboxylic acid group was the essential feature necessary for reaction with the enzyme.

As the enzyme is now commercially available, more reports on its specificity toward different substrates have appeared in the literature and some of these are summarized in Table III.

In their original work, Houmard and Drapeau noted that under certain circumstances aspartyl bonds might be hydrolysed if the adjacent residue was of small molecular size and in fact they found that the –Asp–Gly– bond was slowly hydrolysed, with the explanation that such reaction was a consequence of the β-carboxyl of Asp becoming accessible to the enzyme-binding site; however, it is difficult on this basis to account for some of the cleavages reported in Table III.

Table III. Aspecific cleavages observed using the enzyme *Staphylococcus aureus* protease

Protein	Cleavages observed		Reference
α-subunit of	174–175	Asp-Ala	Ovchinnikov *et al.*
DNA-dependent	197–198	Asp-Leu	(1977)
RNA-polymerase	233–234	Asp-Leu	
	280–281	Asp-Leu	
	21–22	Ser-Thr	
	49–50	Ser-Ser	
	266–267	Ser-Ala	
	313–314	Ser-Leu	
Pike whale RNase	14–15	Asp-	Emmens *et al.* (1976)
	53–54	Asp-	
	83–84	Asp-	
	121–122	Asp-	
	45–46	Thr-Phe	
Kangaroo RNase	121–122	Asp-	
Horse pancreas	7–8	Ser-Met	Evenberg *et al.* (1977)
phospholipase A$_2$	107–108	Ser-Lys	
Ox phospholipase A$_2$	10–11	Lys-Cys	Fleer *et al.* (1978)
	43–44	Arg-Cys	
α-subunit of pea lectin		Asp-	Richardson *et al.* (1978)
		Asp-	

Sutton *et al.* (1977) confirmed the supposition that the –Glu–Pro– bond is not hydrolysed by this enzyme, and also found that N-terminal glutamyl residues were quantitatively released, contrary to the findings of Hitz *et al.* (1977). However, not all glutamyl bonds are susceptible to hydrolysis, strongly suggesting that the local environment has some importance. Vandekerckhove and van Montagu (1977), found that due to its hydrophobic character, the A-protein of Coliphage MS2 was not attacked at all by this enzyme even after prolonged incubation. Hitz *et al.* (1977) reported that glutamyl residues in the sequence –Glu–Glu– were not cleaved. However, other glutamyl residues may resist hydrolysis; Brosius and Arfsten (1978) found that the glutamyl residue at position 2 of the peptide Gln–Glu–Gln–Met–Lys–Gln–Asp–Val–Pro–Ser–Phe–Arg–Pro–Gly–Asp–Thr–Val–Glu, isolated from the ribosomal protein L19, was also not hydrolysed. Nevertheless, this enzyme, in the short time it has been available, has proved extremely valuable to protein chemists: van Dijk *et al.* (1976) in their work on the sequence of muskrat ribonuclease found it useful in the assignment of amide groups; the tryptic peptide isolated (Figure II) was hydrolysed to give two peptides which were purified and analysed clearly indicating where the amide group was located. This enzyme has also been used to help correct the amino acid sequence of the snake

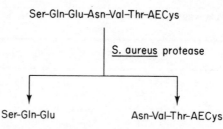

Figure II   Treatment of a tryptic peptide
from muskrat ribonuclease with *S. aureus*
protease

erabutoxins (Maeda and Tamiya 1977). A tryptic peptide was isolated from
these neurotoxins and its partial sequence –Gly–(Glu,Ser)–Ser– deduced;
further work using the subtractive Edman degradation led to the incorrect
formulation –Gly–Ser–Glu–Ser– due to contamination by Glu, arising from
incomplete reaction; however, the correct sequence –Gly–Glu–Ser–Ser– was
found after hydrolysis of the peptide with *S.aureus* protease.

A more recent application of this valuable new proteinase has been by
Begg *et al*. (1978) in their investigation of the location of disulphide bridges
in human β-thromboglobulin, and they conclude that as disulphide
interchange reactions are minimal at its optimal pH (4.0), this enzyme might
become a useful alternative to pepsin in such investigations.

CHAPTER 2

Chemical Methods of Cleavage

(1) CYANOGEN BROMIDE

The pseudo-halogen cyanogen bromide is the most useful of the many chemical methods of peptide chain cleavage that have been introduced. Its discovery by Gross and Witkop (1962) followed a detailed and systematic search for a reagent that would form a sulphonium salt with methionine sufficiently labile to undergo intramolecular displacement and decomposition under relatively mild conditions. This reagent has the unique specificity of breaking peptide bonds on the carboxyl side of methionine residues, and as these residues are relatively rare in proteins, the products formed are usually large peptides. These peptides in the past have been valuable in the ordering of the tryptic peptides obtained from a primary digest of a protein and at present are equally valuable in providing large fragments to subject to automated Edman degradation either in the liquid phase or solid phase sequencer. In the latter procedure they have the added advantage of ease of attachment to the supporting resin as a consequence of their terminal lactone structure.

Mechanism

The mechanism of the reaction of cyanogen bromide with a methionine residue in a peptide chain is illustrated in Figure III. Reaction takes place usually in strong acid solution, at room temperature, and under these conditions methylthiocyanate is formed, and the hydrogen bromide salt of the iminolactone, which subsequently spontaneously decomposes to homoserine lactone.

Specificity

(a) Reaction with Methionine residues

(i) methionine in the middle of a peptide chain

$$\text{--Ala--Met--Phe--} \xrightarrow[\text{CNBr}]{} \text{--Ala--Homoserine lactone} + \text{NH}_2\text{Phe--}$$

19

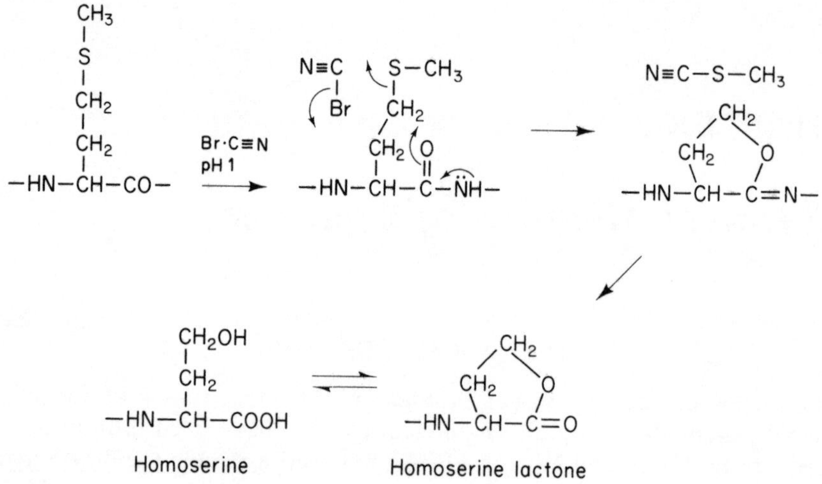

Figure III   Reaction of CNBr with methionine residues in a peptide chain

(ii) methionine N-terminal

$$NH_2Met–Ala– \xrightarrow{CNBr} Homoserine\ lactone + NH_2Ala–$$

(iii) multiple residues of methionine

$$–Phe–Met–Met–Ala– \xrightarrow{CNBr} –Phe–Homoserine\ lactone + Homoserine–$$

$$lactone + NH_2–Ala–$$

(iv) N-acetyl methionine residues

A feature not uncommon is the N-acetyl methionine residue at the N-terminal of a protein; this has been found to incompletely react with CNBr so that approximately 10% of the methionyl peptide bond remains uncleaved due to the formation of O-acetyl-homoserine (Carpenter and Shiigi 1974).

(b) Reaction with Cys, Met(SO) and Met(SO₂) residues

Cysteine is reported to undergo slow reaction with CNBr but peptide chain cleavage is not known to result (Gross and Witkop 1962). However, the reaction undergone does not appear to affect subsequent performic acid oxidation of the cysteine residues, for this is frequently a useful step to render the peptides soluble, particularly if they are large fragments. Methionine sulphone and methionine sulphoxide, occasional artefacts in proteins, do not react with cyanogen bromide and as the former is readily reduced back to methionine during the course of normal acid hydrolysis it could lead to the conclusion being made that incomplete cleavage had occurred; however, a method for the preliminary reduction of these artefacts using mercapto-ethanol under relatively mild conditions has been described recently (Naider and Bohak 1972).

Reaction Conditions

Approximately a 30–100-fold molar excess of CNBr over the methionine present is used*. The solvent most commonly used is 70% (v/v) formic acid and the reaction temperature is usually 25°C, for a period of 24 hours. Following reaction the mixture is diluted 10-fold with distilled water and the solution freeze-dried. Alternatively the reaction may be performed in 70% trifluoroacetic acid at 25°C for 12 hours (Ferrell *et al.* 1978), and is terminated by diluting with distilled water to 5% trifluoroacetic acid concentration. Unreacted protein is precipitated in this way and removed by centrifugation; the soluble peptides are then obtained by lyophilization of the solution.

Lactone/Peptide Conversion

Electrophoresis of freshly isolated cyanogen bromide peptides usually reveals multiple peptides due to the partial hydrolysis of the lactone ring and it is desirable for future handling of the peptides to get them completely either into the lactone or peptide form. Conversion to the lactone is readily accomplished by treating the peptide with trifluoroacetic acid for 1 hour at 25°C. Alternatively the peptide form can be obtained by treating the mixture with a solution of 0.1% NH_4OH for 15 min. However the lactone form is of great value if the peptide is to be subsequently sequenced using the solid phase sequencer, for it can readily be attached to the matrix resin.

Problems

(i) Aspecific cleavages

Some proteins have been found to be difficult to cleave with CNBr: fortunately these are few in number. Scheffer and Beintema (1974) have reported that extensive aspecific cleavage occurred during the CNBr digestion of horse ribonuclease, in particular cleavage on the carboxyl side of several tyrosine residues was found. Apparently this was accompanied by chemical modification of the phenolic groups, as no tyrosine residues could be subsequently identified by specific staining reagents. Further, Simon-Becam *et al.* (1978) found that CNBr digestion of cytochrome C

---

*Cyanogen bromide is highly toxic and as it is quite volatile (m.pt 52°, b.pt 61°) care must be exercised in its use. All handling must be carried out in a fume cupboard and all apparatus after use washed in strong hypochlorite solution, to destroy excess cyanide. Care must be also taken during subsequent freeze-drying to ensure that no volatile cyanides enter the oil pump; an appropriate cold trap should be employed to this end.

from *Schizosaccharomyces pombe* cleaved the peptide chain on the N-terminal side of cysteine residues. This was ascribed to the formation of a dehydroalanyl residue formed during the incomplete S-carboxymethylation of the apoprotein and subsequent cleavage occurring in the acid medium of the reaction, in general, this has been found to facilitate the cyclization of N-terminal glutamine residues forming pyroglutamyl residues.

(ii) Incomplete reaction

Cyanogen bromide digestion of human urogastrone, a potent inhibitor of gastric acid secretion, was found by the I.C.I. investigators to be incomplete (Gregory and Preston 1977); although amino acid analysis of the reaction product indicated complete conversion of the methionine residue to homoserine, no chain fission was found. The amino acid sequence around the particular methionine was –Cys–Met–Tyr–. A similar case was found in the coat protein of the alfalfa mosaic virus (van Beynum *et al.* 1977). Incomplete reaction was found to occur between residues 127–128 (Met–Gln) and 155–156 (Met–Glx) whereas Met–Val bonds in the protein were completely cleaved. These difficulties led to problems during subsequent attempts to purify the cyanogen bromide peptides. It is of interest to note that again the homoserine content of the products was that to be expected for complete conversion of the methionines, yet for some unknown reason the cleavage step had failed. Possibly it is the specific amino acids involved as there are many reports of the sequence Met–Thr presenting difficulties of cleavage. Enfield *et al.* (1975) treated rabbit muscle parvalbumin with CNBr to find that Met residue 2 in the sequence

Ac–Ala–Met–Thr–Glu . . .

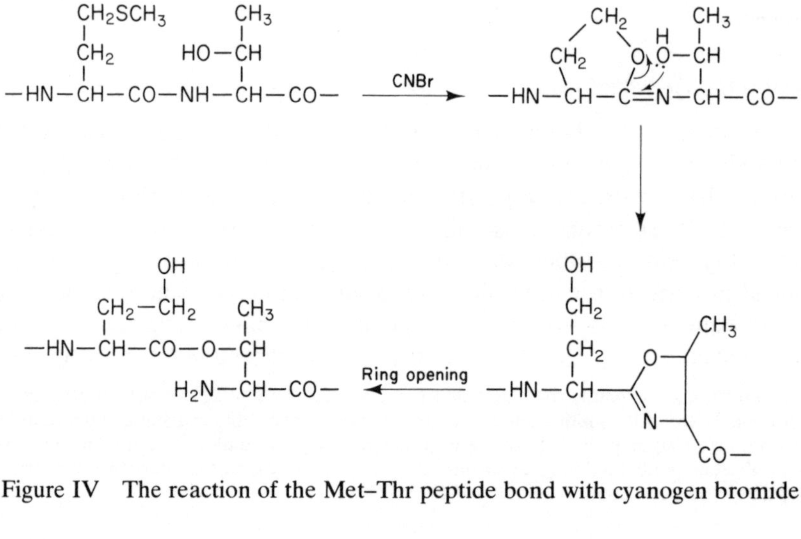

Figure IV   The reaction of the Met–Thr peptide bond with cyanogen bromide

failed to react. Stone and Phillips (1977) also found that a Met–Thr bond (residues 37–38) of dihydrofolate reductase failed to cleave, although conversion to the homoserine had occurred. Possibly this is due to the interaction of the neighbouring hydroxy group of threonine with the intermediate iminolactone to form a homoseryl-O-threonyl peptide as illustrated in Figure IV.

One possible way to circumvent these difficulties is to perform the reaction in 70% TFA rather than the conventional formic acid (Scheffer and Beintema 1974).

(2) HYDROXYLAMINE

Deselnicu *et al.* (1973) have shown, using the model compound Z–Asn–Gly–OEt, that NH_2OH reacts and cleaves specifically the peptide bond between asparagine and glycine amino acid residues. They have suggested that reaction proceeds through the intermediate formation of a succinimide ring and leads eventually to specific fission of the Asn–Gly bond. The reaction mechanism is illustrated in Figure V.

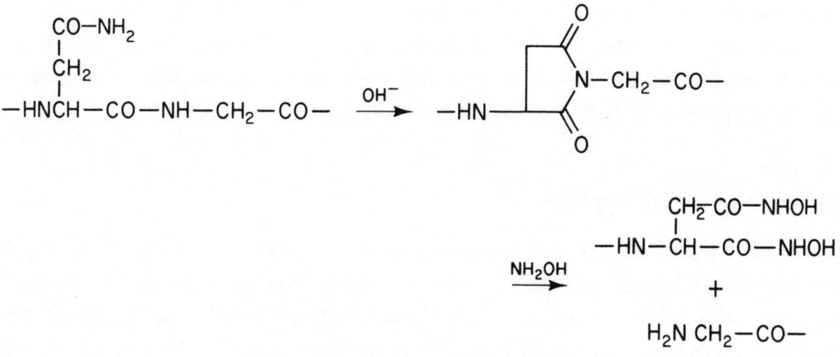

Figure V   The reaction of NH_2OH with the peptide bond between Asn–Gly

This reaction has been used by the Russian biochemists to cleave the single Asn–Gly peptide bond in the α-subunit of DNA-dependent RNA polymerase from *E. coli* (Ovchinnikov *et al.* 1977). Two peptides were obtained after chromatography on Sephadex G100, one from the N-terminal portion contained residues 1–208, and the other was from the C-terminal region and contained residues 209–329 (Figure VI).

This latter peptide had N-terminal glycine and was sequenced up to residue 234 in a Beckman 890C liquid phase sequencer. This information provided vital confirmation of the amino acid sequence of a difficult region of the protein.

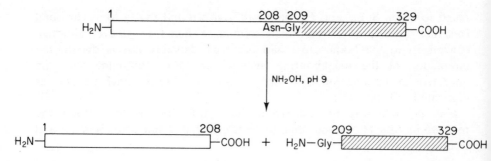

Figure VI   The cleavage of the α-subunit of DNA-dependent RNA-polymerase by
NH_2OH

Reaction Conditions

In a typical reaction the protein is added to a solution of 6 M guanidine HCl containing about a hundredfold molar excess of $NH_2OH.HCl$, previously adjusted to pH 9.0 with 4.5 M LiOH. The reaction is allowed to proceed at 45°C for 4–5 hours, the pH being readjusted every 15 min by the addition of 5 μl portions of 4.5 M LiOH (Frank et al. 1978). Finally the reaction is terminated by lowering the pH to between 2 and 3 using 9% HCO_2H. The reaction mixture is desalted on a small column of Sephadex G25 and the resulting peptides further purified on a column of Sephadex G75.

Alignment of CNBr Peptides

The present trend on protein sequence determination studies is to produce a small number of large peptide fragments. This may be achieved in either of two ways: first the action of proteolytic enzymes might be reduced by a preliminary chemical modification of the protein under study; or secondly a chemical means for chain fragmentation may be used, which may have a greater specificity. In a study of the β-subunit of C-Phycocyanin from the cyanobacterium *Mastigocladus laminosus*, NH_2OH has been used to align the CNBr peptides (Frank et al. 1978). The β-subunit of the protein was treated with NH_2OH and the products separated on a column of Sephadex G75. Six peptides were obtained and following a detailed examination of one of these the alignment of the CNBr peptides could be achieved. A similar strategy was used by Hogg and Hermodson (1977), in their investigation of the structure of the L-arabinose-binding protein from *E. coli*. This protein was treated with NH_2OH and cleavage of the Asn–Gly peptide bond between residues 232–233 was achieved. The resulting peptides were fractionated on Sephadex G75 and from subsequent studies on these peptides it was possible to align two of the cyanogen bromide peptides, which was an essential step in the structure elucidation.

(3) BNPS-SKATOLE

BNPS-skatole* (2-(2-Nitrophenyl sulphenyl)-3-methyl-3'-bromoindolenine), is formed by the reaction of 2-(2-nitrophenyl sulphenyl)-3-methyl indole with one equivalent of N-bromosuccinimide in dilute acetic acid. It is an extremely mild source of positive bromine. Unlike N-bromosuccinimide, this

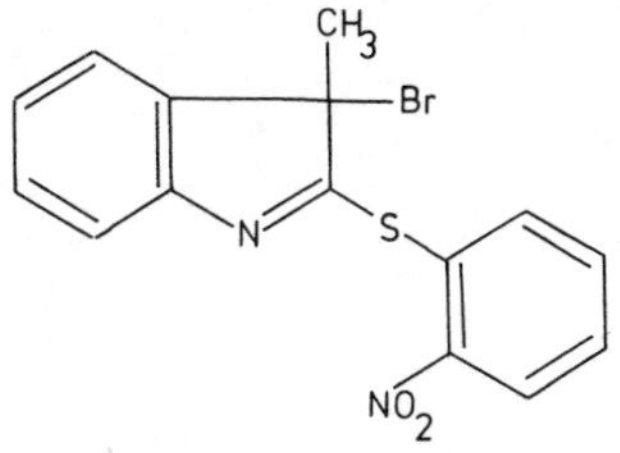

reagent under suitable conditions reacts specifically with tryptophan residues in proteins, and as tryptophan is a rare amino acid in all proteins, the use of this reagent in producing large peptide fragments has increased in favour over the last few years.

General Reaction Conditions

There have been various methods for its use described in the literature and it is obvious that reaction will depend to a great extent on the protein being investigated. In general it is best to do several trial experiments to find the most suitable conditions. A general outline of the procedure used is as follows.

The protein is dissolved in either aqueous acetic or formic acid (50–70%) and to this is added a 5–10-fold excess of BNPS-skatole. Reaction is allowed to proceed at room temperature, in the dark, for between 6 and 48 hours. Following reaction the mixture is diluted with water, and excess reagent/side products removed either by centrifugation or solvent extraction. The reaction mixture is then desalted in a small column of Sephadex G25 before final purification of the peptides is carried out.

Applications and Problems

Protein S_6 of the small ribosomal subunit of *E. coli* was found to contain a single tryptophan residue and BNPS-skatole was used to cleave at this position in an attempt to obtain large peptide fragments suitable for

---

*BNPS-skatole (mol.wt 363.17; m.pt 97–100°C) is unstable at room temperature and decomposes with the liberation of bromine, becoming dark brown in colour. However, if kept in the deep-freeze it is stable for several months. It is not very soluble in aqueous solutions.

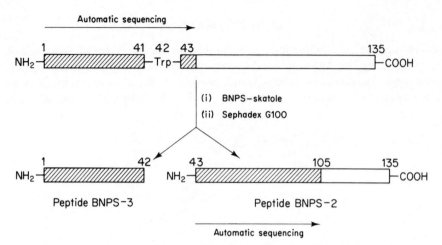

Figure VII   The reaction of BNPS-skatole with ribosomal protein S6

automatic sequencing (Hitz *et al.* 1977). The reaction course was monitored by polyacrylamide gel electrophoresis, which initially indicated that the extent of cleavage was about 16%. However, by carefully adding higher amounts of reagent and improving its solubilisation by adding 25% ethanol it was possible to increase the cleavage up to 50%. The two resulting peptide fragments were obtained by chromatography on Sephadex G100. The largest peptide (BNPS-2) contained 93 residues and the smaller one (BNPS-3) contained 42 amino acids (Figure VII). The amino acid sequence of the first 62 amino acids of peptide BNPS-2 was obtained by liquid phase automatic sequencing, and as the intact protein had been sequenced up to residue 44, by the same method, 82% of the whole sequence had been deduced as a result of this single specific fragmentation.

Cleavage of the single tryptophan residue of the α-subunit of DNA-dependent RNA polymerase was also obtained using this procedure (Ovchinnikov *et al.* 1977), enabling the C-terminal sequence of this protein to be elucidated. It is interesting to note that in this instance the peptide bond cleaved was Trp–Pro, indicating that the imino acid did not hinder the course of reaction. Nevertheless, if proper care is not exercised in this reaction, side reactions may occur leading to the production of a large number of peptide fragments, probably as a result of cleavage at modified tyrosine residues, as was found in the studies on the L-arabinose-binding protein from *E.coli* (Hogg and Hermodson 1977). In this study many more fragments were obtained than expected from the tryptophan content of the protein, and in fact bromotyrosine was subsequently found in some of the peptides studied. However, such side reactions can be reduced and eliminated if the procedure is carried out in the presence of a large excess of free tyrosine, which functions as a scavenger (Fontana 1972). This technique was used by Sepulveda *et al.* (1975), in their studies on pig pepsin. Alternatively, phenol has also been successfully employed to the same end by Frank *et al.* (1978) in studies on the structure of the α-subunit of C-Phycocyanin.

When reaction at tyrosine has been subdued in this way the only other amino acid residue that may undergo oxidation has been found to be methionine, which is converted to the suphoxide. However, this modification can be readily removed at the end of the reaction by the addition of either thioglycollic acid (Fontana 1972), or β-mercaptoethanol as used by Wang *et al.* (1977) in their studies on the amino acid sequence of apomyoglobin from the Pacific dolphin. Therefore, under rigorously controlled conditions only the amino acid tryptophan is modified by BNPS-skatole, and in fact the location of the single tryptophan, at position 128 of the α-subunit of C-Phycocyanin, was established on the basis of cleavage at that point with this reagent (Frank *et al.* 1978).

(4) N-BROMOSUCCINIMIDE (NBS)

This reagent* has been used to cleave peptide bonds on the carboxyl side of tyrosine and tryptophan amino acid residues in proteins and peptides. Its selectivity is, however, less than that of BNPS-skatole. It is a source of positive bromine and it is this reactive species that is responsible for polypeptide chain cleavage. The positive bromine ion reacts with the indole nucleus of tryptophan, oxidizing it and inducing peptide fission with the formation of a lactone structure at the C-terminal of the newly formed cleaved peptide. Oxidation sometimes occurs without peptide chain cleavage, depending upon the reaction conditions, which must be experimented with to maximum advantage.

General Procedure

The protein is dissolved in either a solution of 8 M urea adjusted to pH 4.0 with acetic acid, or an aqueous solution of 50% acetic acid. At least a

---

*N-bromosuccinimide (mol.wt 178; m.pt 173°C) is best purified before it is used for peptide chain cleavage by recrystallisation from ten times its weight of water (Fieser and Fieser 1967).

twofold molar excess of NBS is added in a small volume of the same buffer. Reaction is allowed to proceed for about one hour, at room temperature, in the dark. Immediately after reaction the mixture is fractionated on a column of Sephadex G50.

Applications and Problems

The method has been successfully applied to the elucidation of the structure of the coat protein of alfalfa mosaic virus (van Beynum *et al.* 1977). In this study a cyanogen bromide peptide CBII was treated with NBS and two peptides isolated by chromatography on Sephadex G50, CBII-BS1 with nine residues, and CBII-BS2 with 26 residues. Studies on the amino acid sequence of these peptides provided additional information for the correct alignment of the tryptic peptides of this cyanogen bromide peptide. However, there are problems associated with this method of cleavage, mainly because of its lack of specificity, as illustrated by the investigations on the structure of a cyanogen bromide peptide (CNBr VI) from the enzyme dihydrofolate reductase from a methotrexate-resistant mutant of *E.coli* (Bennett *et al.* 1978). The carboxymethylated peptide was treated with NBS for 30 min in 50% acetic acid and the products fractionated on Sephadex G50. It was clear from the number of peaks obtained that cleavage had occurred at both tryptophan and tyrosine residues. However, from the mixture only two peptides were isolated and studied, *N*1 (residues 93–133) and *N*2 (residues 134–159) (Figure VIII). Edman degradation of peptide *N*2 provided essential evidence for the sequence of this cyanogen bromide peptide.

Attempts have been made to limit the cleavage of proteins by NBS to just the tyrosine residues by prior chemical modification of the tryptophan. This

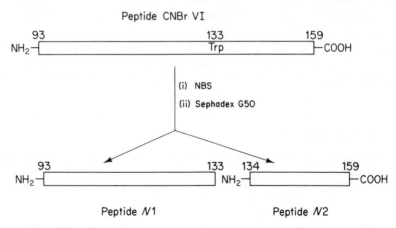

Figure VIII Cleavage at the single tryptophan residue of peptide CNBrVI by NBS (other peptides were also obtained, probably due to cleavage at tyrosine residues)

has been accomplished by ozonolysis of the indole nucleus to give N-formyl-kynurenine derivatives, which are unreactive towards NBS; this technique has been successfully applied to studies on the α-chain of human haemoglobin (Previero et al. 1967).

Recent studies, however, have shown that if N-chlorosuccinimide (NCS) is used in place of NBS, many of the problems are overcome and specific cleavage at tryptophanyl bonds in proteins is obtained (Shechter et al. 1976) and it would seem that this reagent is comparable to BNPS-skatole in its selectivity.

(5) PARTIAL ACID HYDROLYSIS

Partridge and Davis (1950) observed that partial acid hydrolysis of a protein with dilute (0.25 M) acetic acid, under reflux, led to the preferential release of aspartic acid. The reaction may be formulated as:

$$\ldots \text{X–Y–Asp–W–Z} \ldots \xrightarrow{\text{H}^+} \ldots \text{X–Y} + \text{Asp} + \text{W–Z} \ldots$$

where X, Y, W and Z represent any amino acid, other than Asp and Asn. Schultz et al. (1962) showed that if dilute mineral acid is used then the Asn is hydrolysed initially to Asp, and is subsequently released.

This method of selective cleavage at aspartic acid residues of a protein chain has been successfully exploited by Croft (1972a, b) in the elucidation of the amino acid sequence of the lens protein γ-crystallin. In this study a preliminary investigation of the release of aspartic acid from the protein by boiling 0.25 M acetic acid was made (Figure IX). The protein (25 mg) was then hydrolysed, in 20 ml of 0.25 M acetic acid at 100° in an evacuated

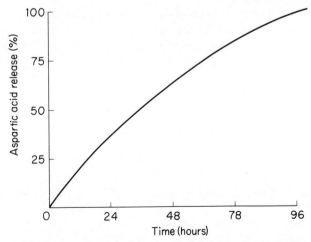

Figure IX The release of aspartic acid from γ-crystallin
by hydrolysis with 0.25 M aspartic acid at 100°C

Carius tube, and after 86 hours the solution was freeze-dried. By using a combination of electrophoresis and paper chromatographic methods the peptide Ser–Ile–Arg–Val was isolated, which provided important information for the alignment of two tryptic peptides.

Other workers have also resorted to partial acid hydrolysis to provide them with essential overlapping information. Fowler and Zabin (1978), in their monumental studies of the amino acid sequence of β-galactosidase from *Escherichia coli*, used a novel method of partial acid hydrolysis. The carboxymethylated enzyme (0.5 g) was incubated at pH 2.5 in 7 M guanidine hydrochloride for 120 hours at 40°. Extensive cleavage throughout the 1021-residue molecule was observed and from the complex mixture of peptides was isolated peptide AS-1 which provided the key information for the alignment of two cyanogen bromide peptides.

(6) N → O ACYL REARRANGEMENT

In anhydrous acid the peptide bond of seryl (and to a lesser extent threonyl) is known to undergo intramolecular rearrangement to the adjacent hydroxyl group to form an ester link (Elliott 1952), thus:

$$-CHR \cdot CO \cdot NH \cdot CH \cdot CO \cdot NH- \quad \xrightleftharpoons[\ OH^-\ ]{\ H^+\ } \quad \overset{+}{H_3}N-CH \cdot CO \cdot NH-$$
$$HO \cdot CH_2 \qquad\qquad\qquad\qquad -CHR \cdot CO \cdot O \cdot CH_2$$

The ester bond formed is much more labile to nucleophilic attack than the original peptide bond and under mildly basic conditions may be selectively and completely cleaved. However, as under these mildly basic conditions the N → O acyl shift is reversible, the amino groups must first be blocked before selective cleavage is attempted.

This procedure has been successfully applied by Macleod *et al.* (1977) to histone H1 from trout testis. In this investigation the histone was incubated in concentrated H_2SO_4 for an extended period of time and before cleavage at the ester bonds was attempted, the exposed amino groups were blocked with a reversible acylating agent. The ester bonds were then cleaved with the powerful nucleophile hydroxylamine, thus:

$$-H \cdot N-CH \cdot CO \cdot NH- \quad \xrightarrow{\ NH_2OH\ } \quad -HN \cdot CH \cdot CO \cdot NH-$$
$$-CHR \cdot CO \cdot O \cdot CH_2 \qquad\qquad HO \cdot CH_2$$
$$\qquad\qquad\qquad\qquad -CHR \cdot CO \cdot NH \cdot OH$$

and in this way three peptides were obtained, each containing N-terminal serine.

CHAPTER 3

The Purification of Peptides

One of the major problems in sequence analysis is the fractionation and purification of peptides. Unfortunately, during the last few years there has been little improvement in the techniques used. The strategy employed by the investigator may be the classical approach, that is the protein is fragmented into a large number of small peptides that have to be separated and purified; or the protein is split into a small number of large peptides. There are snags in both procedures: in the first case the fractionation is difficult because of the large number of fragments present and in the second instance the peptides, because of their size, are usually difficult to solubilize, and the use of disaggregating media, such as 6 M urea, may have to be employed, which is both messy and expensive.

The main procedures in current usage for the purification of peptides are:

(1) gel filtration;
(2) ion-exchange chromatography;
(3) preparative chromatography and electrophoresis.

(1) GEL FILTRATION
(gel chromatography, gel permeation chromatography etc.)

Gel filtration finds its main use in peptide purification either as a first step in the fractionation of an enzyme digest, or in the separation of a small number of peptides differing considerably in molecular weight, for example, after treatment of a protein with cyanogen bromide. The principles of gel filtration are very simple. If a sample consisting of substances differing in molecular size is applied to a column of gel filtration media and is eluted with a suitable buffer, the sample percolates through the bed. If the molecules are too large to penetrate the matrix they are eluted ahead of those that can, the latter being eluted in order of decreasing molecular weight. However, peptides of similar molecular weight may be separated, in the presence of detergents, such as dodecylamine, or dodecanoic acid, on the basis of charge differences (Strid 1973).

It is now generally appreciated that phenomena other than molecular sieving occur during gel chromatography of peptides. These include:

(*a*) interaction between acidic peptides and carboxyl groups on the gel

31

resulting in their partial exclusion; this interaction may be reduced by using weak buffers;

(b) aromatic amino acids interact with the gel matrix and are retarded; this is sometimes dramatic with tryptophanyl peptides;

(c) hydrophobic peptides may participate in hydrophobic bonding with the gel matrix.

These interactions are never very significant except in the case of tryptophanyl peptides, when they can be used to advantage in their purification. Gel filtration therefore offers a simple, reproducible (with 100% recoveries) and quick method for peptide purification.

There are two main types of gel matrix employed in peptide purification, namely Sephadex and Bio Gel.

Sephadex

This is prepared by cross-linking dextran (poly(α-1,6-D-anhydrogluco-pyranose)) with epichlorohydrin. The extent of cross-linking is controlled by the dextran chain length and the percentage of epichlorohydrin. The types of gel available, together with their fractionation ranges, are given in Table IV. Sephadexes are insoluble in water yet stable in weakly acid or alkaline solution; however, above pH 12 and below pH 2 they are degraded, and as small amounts of soluble dextrans are to be found in the effluents of these gels it is inadvisable to use them in the purification of glycopeptides.

Table IV. Gel filtration media suitable
for peptide purification

Type	Fractionation range (molecular weight)
(i) Sephadex	
G10	< 700
G15	<1500
G25	1000–5000
G50	1500–30 000
(2) Bio-gels	
P2	200–1800
P4	800–4000
P6	1000–6000
P10	1500–20 000

Bio-Gels

These are prepared by cross-linking acrylamide (CH_2=CH.CONH$_2$) with N,N'methylene bis acrylamide (CH_2=CH.CONH.CH$_2$NH.CO.CH=CH$_2$). These gels are stable in most commonly used buffers in the pH range 1–10; however, beyond pH 10 the amide groups of the matrix become labile and tend to hydrolyse.

Procedures

Equipment

The column used is best made of sturdy plastic or glass (not metal), with a bed support of the same diameter as the column, made of nylon cloth (400 mesh) and not of glass sinters which tend to get clogged, or glass wool which introduces an undesirable dead volume. The outlet tube should be of narrow bore and if possible the column should be enclosed in a constant-temperature water jacket. The length : diameter ratio of the column should be about 100 : 1; a greater length may be obtained by coupling several columns in series. Gregoire and Rochat (1977) obtained an excellent separation of peptides from the tryptic digest of a neurotoxin from the African cobra by using two 2 × 200 cm columns in series; and Bhown *et al.* (1977) achieved a good separation of the tryptic peptides from human J-chain by using an arrangement of three columns, 2 × 112 cm each in tandem (Figure X).

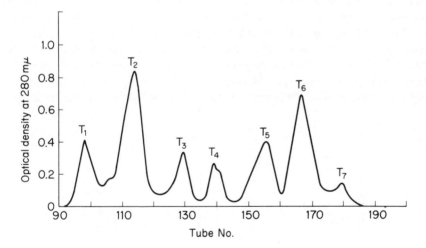

Figure X   Gel filtration of tryptic digest of citraconylated J-chain (160 mg) on two P-10 columns (Bio–Rad, 200–400 mesh, 2.0 × 112 cm) in 1% NH_4HCO_3 attached in tandem. (Reprinted with permission from *Bhown et al.* (1977), *Biochemistry*, **16**, 3501. Copyright by the American Chemical Society)

The sample to be applied to the column must be free of suspended matter and have a viscosity not greater than twice that of the eluant buffer. The sample volume should be about 1% of the bed volume. If a pump is employed the flow rate should not exceed 90% that obtainable by gravity flow. The resolution is best with a low flow rate and no advantage is gained by using ascending flow rather than the usual descending method.

Elution Solvents

The first criterion for a solvent is that it will dissolve adequately the peptides; second, that it will not interfere with the detection of the peptides; and third, it must be possible to remove it without contaminating the peptide with salts or other residues. The following have been successfully employed for the separation of peptides on gel chromatography:

  (i) 0.1 M ammonium acetate, pH 8–9;
 (ii) 1% NH₄HCO₃;
(iii) 1 M NH₄OH, or 1 M acetic acid;
 (iv) 6 M guanidine hydrocholoride, or 6 M urea;
  (v) 50% acetic acid, or formic acid.

The last two have been used to fractionate cyanogen bromide peptides and other large fragments. With 6 M urea the danger of carbamoylation is always present, particularly if old or heated solutions are used; these should be avoided; after fractionation the urea can be removed, provided the peptides are of high molecular weight by dialysis. The same applies for 6 M guanidinium chloride. Strong solutions of acetic and formic acid may be removed by high-vacuum distillation using a two-stage oil pump, protected with two liquid nitrogen traps and a third trap of sodium hydroxide pellets. Figure XI shows the successful separation of the cyanogen bromide peptides from dihydrofolate reductase by gel filtration on a column of Sephadex G50 eluting with 50% acetic acid (Bennett *et al*. 1978). As gel chromatography does not differentiate between the acid/lactone forms of the cyanogen

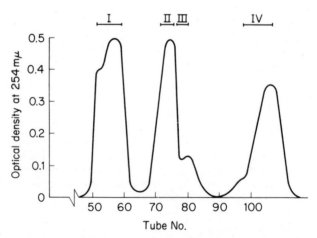

Figure XI Gel filtration of peptides produced by CNBr digestion of dihydrofolate reductase. The column (0.9 × 100 cm) was eluted at the rate of 10 ml/h with 50% acetic acid. (Reprinted with permission from Bennett *et al.* (1978), *Biochemistry*, **17**, 1328. Copyright by the American Chemical Society)

bromide peptides, it has an inherent advantage over other forms of purification.

Detection of Peptides from Gel Filtration Columns

The usual method for detecting peptides in the eluent is by absorption in the UV at 280 mμ, 230 mμ, or 215 mμ; however, if the absorption of the elution buffer, or the lack of specific absorption in the peptide, makes this impracticable then peptides may be detected with ninhydrin, after alkaline hydrolysis of a small amount of each fraction. Alternatively, reaction with fluorescamine is more direct and readily performed (e.g. 10 μl of each tube is spotted on chromatography paper, previously sprayed with 0.4 M sodium borate, pH 9.3, and then dipped in 0.1% fluorescamine in acetone); this method has been used in many instances.

(2) ION-EXCHANGE CHROMATOGRAPHY

Soon after the introduction of ion-exchange chromatography for amino acid separations, attempts were made to use these systems for the separation of peptides from enzyme digests. Unfortunately, as inorganic salts were used in the construction of the pH gradients, great difficulty was experienced in the extrication of the small quantities of peptide material from the bulk of inorganic salts. Various ingenious methods were suggested to overcome these difficulties, none of them too successful, and it was not until volatile buffers were introduced by S. G. Waley and others, that the use of ion-exchange chromatography for the purification of peptides became a practicable proposition.

Table V. Ion exchange media for peptide fractionation

Type	Functional group	Properties
(a) CATION		
Dowex 50	$-SO_3^- Na^+$	strongly acidic
(equivalent BioRad		strong cation exchanger
Aminex 50W)		weak cation exchanger
(CM-Cellulose		
CM-Sephadex)	$-CH_2CO_2^- Na^+$	
P-Cellulose	$-O-P{=}O \ 2Na^+$ (with $O-$ groups)	intermediate strength cation exchanger
(b) ANION		
Dowex 1	$-\overset{+}{N}(CH_3)_3Cl^-$	strongly basic, strong anion exchanger
(DEAE-Cellulose		weak anion exchanger
DEAE-Sephadex)	$-CH_2CH_2\overset{+}{N}H(C_2H_5)_2Cl^-$	

Some of the ion-exchange resins in common use in peptide purification at the present time are listed in Table V.

Separation of Peptides on Bio-Rad Aminex 50W × 2

Only the minimum of details of the method will be given here, as excellent technical reviews are available (Schroeder 1972). Various column sizes may be used, and the ones illustrated (Table VI) are those in common use in the author's laboratory. Excellent separations are obtained using a linear gradient of buffers that gradually increase in concentration and pH (Table

Table VI. Particulars of columns and eluents suitable for peptide fractionation

Column dimensions (cm)	Weight of digest (mg)	Volume applied (ml)	Volume of eluents (ml)	
			pH 3.1	pH 5.0
55 × 0.9	90	1	200	200
90 × 0.9	200	2	500	500

VII) this being achieved by employing two vessels of matching diameter connected together (Figure XII).

The most useful and universal method for the fractionation of tryptic digests (and it has been these digests more than any other that have been studied in this way) is the strongly acidic sulphonated polystyrene resin, Dowex 50, or its equivalent Bio-Rad Aminex 50W. The degree of cross-linkage (which is usually expressed as × 2 etc.) represents the porosity of the resin to molecules, and has been usually 2% (i.e. × 2); however, satisfactory results can be obtained using the higher cross-linked resins up to × 8 (which is the resin normally employed in amino acid analysis). Consideration as to the mesh size should be given as this determines not only the flow rate but the resolution obtained (a satisfactory one is the 200–400 mesh size).

Before any attempt is made to fractionate a complex peptide mixture, it is essential to do a preliminary analysis to estimate the total number of

Table VII. Volatile buffers for ion-exchange chromatography (Schroeder, 1972)

Pyridine concentration (M)	pH	Pyridine	Glacial acetic acid
		(ml/2000 ml)	
0.2	3.1	32.3	557
2	5.0	322.5	286.5

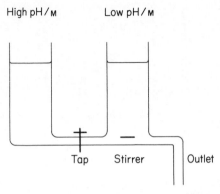

High pH/M   Low pH/M

Tap   Stirrer   Outlet

Figure   XII Apparatus   for   linear
gradient elution

peptides in the digest. This is best achieved by peptide mapping, and as the peptides are eluted from the column they are checked off against the original composition of the mixture, so as to ensure that no peptides are lost. In particular, basic peptides may on occasions be irreversibly bound to the polystyrene resins, as was found by Ovchinnikov *et al.* (1977) who 'lost' two strongly basic peptides from a *S. aureus* protease digest of the α-subunit of RNA polymerase, when fractionated on the AG50 × 2 resin. This simple system of 'book-keeping' is an easy way of insuring against such a loss, that could have disastrous consequences when the protein sequence is reconstructed.

The resin to be used is packed into the water-jacketed column under gravity and equilibrated with starting buffer overnight. The tryptic digest is brought to pH 2 with 6 M HCl and any insoluble material removed by centrifugation. This material is known colloquially as the 'tryptic core' and must be preserved for subsequent examination. The clear supernatant is applied to the column under force of gravity and elution is commenced with the linear gradient from 0.2 M pyridine–acetate (pH 3.1) to 2 M pyridine–acetate (pH 5.0), the volumes employed depending on the column size. The temperature of elution is 40° and fractions of approximately 2 ml are collected every 20 min. Peptide material may be detected in the fractions after alkaline hydrolysis of a small portion and reaction with ninhydrin. Excellent results are almost always obtained: Figure XIII illustrates the separation of tryptic peptides from γ-crystallin fraction IV (Slingsby and Croft 1978). Peptides are readily recoverable from the main peptide peaks after removal of the volatile solvents by rotary evaporation and lyophilization. Occasionally some pyridine salts persist, with the result that streaking is observed on subsequent paper chromatography and the only recourse is to repeat the lyophilization process. Yields of about 70% recovery are normally obtained—not as high as from gel chromatography—however, the resolution obtained is far superior.

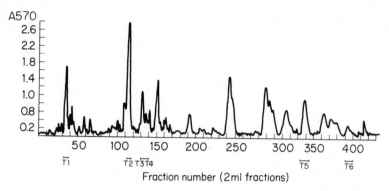

Figure XIII Fractionation of tryptic peptides from γ-crystallin on Bio–Rad Aminex 50 W × 2, column size 90 × 0.9 cm. Elution, at 40°C, was achieved with a gradient from 0.2 M pyridine/acetate buffer pH 3.1 to 2 M pyridine acetate buffer pH 5.0, 500 ml of each buffer being used. (Reproduced from Slingsby and Croft 1978)

Many workers have reported that occasionally the same peptide may be found in two separate parts of the resulting chromatogram. In some cases it is readily explained, if for example methionine was present in the peptide, the two forms probably represent different oxidation levels of the methionine sulphur. However, some instances are not so easily accounted for; thus Slingsby (1974) found the peptide

Ala–Val–Asp–Phe–Tyr

in two well-separated fractions during ion-exchange chromatography of a tryptic digest of γ-crystallin (fraction IV), and it can only be speculated that this peptide must exist in two distinct conformers that are resolved due to differ interactions with the polystyrene matrix. The presence of two adjacent aromatic residues could be a source of such conformational disparity.

The Use of Micro-Columns

One of the problems associated with polystyrene resins is that high losses of certain peptides may occasionally occur. In the main these are due to both ionic and non-ionic interactions, and may be reduced, although not altogether eliminated, by the use of micro-columns containing only a small quantity of ion-exchange resin. Such a novel column of dimensions 0.3 × 10 cm was used by Wittmann-Liebold and Marzinzig (1977) to resolve the tryptic peptides from the ribosomal protein L28, with Dowex M71 resin (equivalent to 50W × 7) and eluting with a pyridine–formate gradient at 55°C. An even smaller column (0.25 × 10 cm) was used by Chen (1977) to separate the tryptic peptides from 2 mg of the ribosomal protein S9. However, care must be taken when using these micro-columns that peptides are not lost by adherence to the glass surfaces, as was discovered by Chen et al. (1975) when a tryptic peptide of 14 amino acid residues was lost in this way from a tryptic digest of the ribosomal protein L27. The technique generally employed to detect peptides from these micro-columns is TLC. A small sample (40 μl) is removed from each fraction and spotted on a thin layer plate of cellulose and chromatography is performed in the solvent pyridine–n-butanol–acetic acid–water (50 : 75 : 15 : 60, v/v). The peptides are detected by staining with ninhydrin.

Other Ion-Exchange Media

Successful separations of peptides are not always achieved on polystyrene resins: Jackson et al. (1977) attempted to fractionate proteolytic digests of hen plasma low density lipoproteins on Dowex resins and after many unsuccessful attempts resorted to other methods of fractionation. A selection of useful alternative ion-exchange materials is given in Table VIII. In general the resolution obtained with these materials is inferior to that achieved with Dowex resins, but the loading capacity is greater (Chen 1977).

By a skilful combination of these various techniques the fractionation of the peptides of a tryptic digest is now readily achieved and this can be no better illustrated than by the work of Chen et al. (1975) on the structure of protein L27 which was successfully sequenced using a total of 24 mg of protein. Two tryptic digestions were performed, one on the citraconylated protein, a total of 10 mg of protein being used. The skilful way the peptides were fractionated is illustrated in Figure XIV.

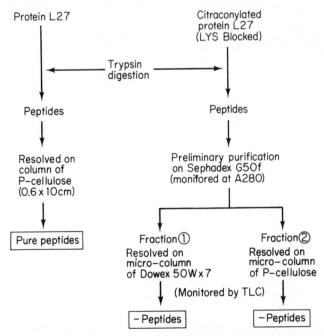

Figure XIV The fractionation of peptides from 10 mg of the ribosomal protein L27 by Chen et al. (1975)

Table VIII. Selected examples of the purification of peptides using different ion-exchange resins

Matrix	Type	Nature of protein digest	Elution	Method of detection	Reference
Polystyrene	Dowex 71 (Beckman resin M71 equivalent to Dowex 50W × 7)	Tryptic digest of ribosomal protein L28	Pyridine–formate gradient 0.1 M (pH 2.7)– 1.0 M (pH 6.5) and 1.0 M (pH 6.5)– 2.0 M (pH 6.5)	Not given	Wittmann-Liebold and Marzinzig (1977)
Cellulose	DEAE	Purification of selected tryptic peptides	Linear gradient 0.025 M–0.5 M NH$_4$HCO$_3$	Fluorescamine	Bhown et al. (1977)
	CM		Linear gradient 0.025 M–0.5 M ammonium acetate (pH 4.5)		
	Phospho-P11	Tryptic peptides from CNBr peptide of chick skin collagen	Linear salt gradient 0 → 0.3 M NaCl, in 0.04 M sodium acetate buffer (pH 3.8)	Absorption at 230 mμ	Dixit et al. (1977a,b)
Sephadex	DEAE–(A25)	Purification of tryptic peptides	Linear gradient 0.005–0.3 M NH$_4$HCO$_3$ pH 8.5	Not given	van Hoegaerden and Strosberg (1978)
	SP–	Separation of CNBr peptides	0.17 M pyridine acetate pH 4.7	Ninhydrin	Ferrell et al. (1978)

(3) PREPARATIVE PEPTIDE MAPPING

The purification of peptides by preparative peptide mapping is both rapid and sensitive. It is also an easy technique to master. In principle, the enzyme digest is subjected to a two-stage process involving a combination of electrophoresis and chromatography, one of these being performed in a direction at right angles to the other. It may be carried out on sheets of chromatography paper, or on thin layers. The former has been widely used in the past, but due to the increased demand for greater sensitivity, thin layer peptide mapping is now in vogue. This extremely simple two-step procedure, that can readily be performed within a 24-hour period, may resolve most of the components of an enzymic digest of a protein. However, to isolate the individual peptides they must be adequately detected, without loss of their chemical integrity, and this demands a mild and simple reagent. Ninhydrin has commonly been used for this purpose, but the new reagent fluorescamine is now considered to be preferable.

Peptide Mapping Using Paper Chromatography

Of the two chromatographic media available, paper is probably the easier to handle and is considerably cheaper than thin layers. Furthermore, advantage may be taken of the many grades of paper available (Table IX). No. 20 paper has been found to result in excellent peptide maps in the author's laboratory (Figure XV), and it is surprising that this paper has not found much use in other laboratories.

For the electrophoretic separation, either high voltage or low voltage may be used. The author has found that the latter results in superior separations, albeit taking longer to perform. Furthermore, high-voltage electrophoresis requires access to expensive equipment. Chromatography, which is usually performed following the electrophoretic separation, is normally carried out

Table IX. Comparison of different qualities of chromatography paper useful for preparative peptide mapping

Grade	Loading capacity (mg)	Resolution	General comments
3MM	5–8	Not very good	Solvent flow is fast in chromatography direction; easy to handle.
No. 1	<1	Better than 3MM	Difficult to handle
No. 20	<2	Excellent resolution	Solvent flow is slow in chromatography direction; easy to handle.

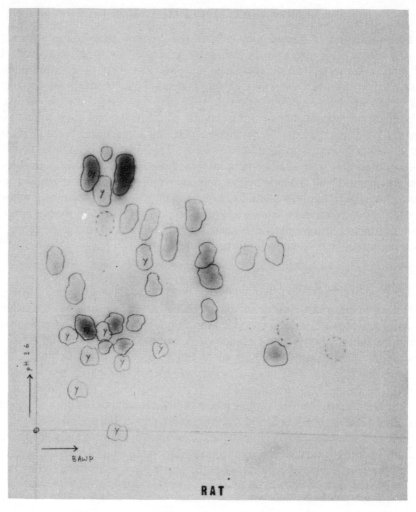

Figure XV   Peptide map of a tryptic digest of rat γ-crystallin on Whatman No. 20 paper. Electrophoresis was performed first at pH 3.6, followed by chromatography at right angles in the solvent BAWP

in the descending mode. The actual practical aspects of these techniques are beyond the scope of this review, and the reader is asked to refer to the many excellent texts available in this field. However a selection of several of the more useful electrophoresis buffers and chromatographic solvents are included for completeness (Table X).

Improved peptide separations may be obtained if the chromatography step is repeated (Hoerman and Kamel 1967).

Table X. Some useful buffers and solvents for peptide mapping on paper

Application	Components (v/v)	pH	Reference
Electrophoresis	pyridine : acetic acid : water; (25 : 1 : 474)	6.5	Sargent (1965)
	pyridine : acetic acid : water; (1 : 10 : 989)	3.6	
	formic acid : acetic acid : water; (52 : 29 : 919)	1.9	
	pyridine : acetic acid : water; (1 : 10 : 289)	3.5	Kondo et al. (1978)
	pyridine : acetic acid : water; (25 : 1 : 225)	6.5	Walker et al. (1977)
	pyridine : acetic acid : water; (3 : 20 : 780)	3.7	Closset et al. (1978)
Chromatography	butan-1-ol : pyridine : acetic acid : water; (15 : 10 : 3 : 12)		Waley and Watson (1953)
	butan-1-ol : acetic acid : water; (4 : 1 : 5)		Evenberg et al. (1977)
	(upper phase only; lower in bottom of tank)		
	pyridine : isoamyl alcohol : 0.1N NH$_4$OH; (6 : 3 : 5)		Tanaka et al. (1977)
	butan-1-ol : pyridine : acetic acid : water; (15 : 10 : 1 : 12)		Walker et al. (1977)

Peptide Mapping Using Thin Layer Chromatography

The use of thin layers for peptide mapping offers the considerable advantages of increased speed and sensitivity. Thin layer plates are also more compact and easier to handle, requiring less bulky equipment. However, they are costly to purchase.

Suitable thin layers of approximately 0.1 mm in thickness may be of silica, polyamide, or cellulose. The latter are the most useful and are available commercially, on plastic, or aluminium foil backings. Chen *et al.* (1975) have recommended that before use the layers should be purified by elution with 5% pyridine, followed by 1.5% formic acid. Again, electrophoresis may be either at high voltage or low voltage, with chromatography, which is usually performed overnight, in the ascending mode. Typical buffers and solvents are shown in Table XI.

The usual load on thin layers is considerably less than on papers, a typical amount being about 0.2 mg. Usually more material is available and it is generally the procedure to run a number of plates simultaneously.

Table XI. Useful buffers and solvents for peptide mapping on thin layers of cellulose

Application	Components (v/v)	Reference
Electrophoresis	pyridine : acetic acid : acetone : water; (1 : 2 : 8 : 40) (pH 4.4)	Chen (1977)
Chromatography	butan-1-ol : pyridine : acetic acid : water; (15 : 10 : 3 : 12)	Chen *et al.* (1975)
	butan-1-ol : pyridine : acetic acid : water; (75 : 50 : 16 : 15)	Chen (1977)

Detection and Elution of Peptides from Paper and Thin Layer Chromatograms

The detection of peptides on maps may be accomplished either by dipping, or spraying, with an appropriate reagent, or by autoradiography, this latter technique being particularly suitable in micro-sequencing.

In the past, a dilute solution of ninhydrin (2 mg/ml in acetone) has been widely used to detect peptides; however, for preparative purposes, it has the drawback that there is considerable destruction of the N-terminal residue. Fluorescamine (shown here) is probably the best reagent at present available to detect peptides, which when employed at the appropriate concentration leads to no loss of the amino-terminal residue. This reagent, which is non-fluorescent, reacts at alkaline pH with amino groups to form a highly fluorescent product. In a typical example Fleer *et al.* (1978) used this reagent to detect the peptides on finger-prints of the protein phospholipase A_2. The paper was first sprayed with 10% (v/v) pyridine in acetone, followed by a solution of 2 mg of fluorescamine in 100 ml 1% (v/v) pyridine in acetone. After drying, the peptides appeared as fluorescent spots, which were cut out and eluted. Vandekerckhove and van Montagu (1974) estimate that this reagent is capable of detecting as little as 0.01 nmol of a peptide. It is also capable of locating peptides having N-terminal isoleucine, which are not always detectable with ninhydrin: however, it does not detect peptides having N-terminal proline residues, unless lysine is present in the peptide.

After detection the peptides may be eluted with any, or a combination of the solvents in Table XII. Beyreuther *et al.* (1975) recommend that this elution be performed at 4°C, so as to avoid any possible deamidation of the peptide.

Table XII. Useful solvents for the elution of peptides from peptide maps

Solvent	Reference
50% acetic acid	Chen (1977)
50% formic acid	Vandekerckhove and van Montagu (1974)
50% pyridine	Evenberg *et al.* (1977)
1% NH_4HCO_3 (pH 9.0)	Evenberg *et al.* (1977);
0.1 M NH_3	van Den Berg *et al.* (1977)
1 M acetic acid	Beyreuther *et al.* (1975)
electrophoresis buffer	van Hoegaerden and Strosberg (1978)

No exact formulation for the successful elution of a peptide can be given, for it depends to a great extent on the properties of the peptide that is to be eluted. Some investigators, for example Fleer *et al.* (1978), to ensure successful elution of the peptide, employ a succession of three solvents. Most peptides would be eluted under these conditions, but there is always the odd hydrophobic peptide that evades all elution attempts. These difficult peptides are not always large, for example the pentapeptide

Ala–Ile–Asp–Leu–Tyr

was found to resist all the usual elution solvents, presumably as a consequence of its hydrophobic character; it was eventually successfully eluted by the imaginative application of the chromatographic solvent (Slingsby 1974).

CHAPTER 4

The Sequenator

The sequenator is the product of the scientific genius of Pehr Edman (1916–1977). In 1950 he published his paper, which now must rank as one of the most important of recent times: '*A Method for the Determination of the Amino Acid Sequence in Peptides*' (Edman 1950). For the remainder of his life Edman was committed to the development of his concept of a fully automatic machine for the sequential degradation of proteins. This vision culminated in the publication, in the first issue of the *European Journal of Biochemistry*, of his article describing such an instrument, to which he gave the name 'sequenator' (Edman and Begg 1967) (see Figure XVI). The

Figure XVI Photograph of the original sequenator. (Reproduced from Edman and Begg (1967) by permission of *European Journal of Biochemistry*)

47

Table XIII. Some of the results obtained using commercial sequencers (excluding the Beckman Sequencer)

Sequencer model	Manufacturer	No. of residues sequenced	Protein/peptide, and the approximate weight used (if given)	Reference
JEOL JAS 47K	JEOLCO (Japan)	40	Ferredoxin (4 mg)	Hase et al. (1976)
		Not given	Cardiotoxin of Bungarus fasciatus venom	Lu and Lo (1978)
SOCOSI P110	SOCOSI (France)	16	Horse MSEL- neurophysin	Chauvet et al. (1977)
		45	Whale MSEL-neurophysin	Chauvet et al. (1978)
SOCOSI PS100	SOCOSI (France)	26	Toxin III from Anemonia sulcata (1 mg)	Martinez et al. (1977)
		46	Neurotoxin I of the snake Naja mossambica mossambica (2 mg)	Gregoire and Rochat (1977)
Illitron	Illinois Tool Works, Chicago (USA)	36	Soyabean leghaemoglobin C_2	Hurrell and Leach (1977)
			κ-chain of mouse myeloma	Smith (1978)
		40	Bence Jones protein Dil (5 mg)	Smithies et al. (1971)

Table XIV. Some results obtained using the Beckman Sequencer

Protein/Peptide	No. of residues sequenced	Approximate amount used (mg)	Average repetitive yield (%)	Reference
C-Phycocyanin (β-)	100	5	Not given	Frank et al. (1978)
C-Phycocyanin (α-)	85	5	Not given	Frank et al. (1978)
Thrombin (β-chain)	70	Not given	98	Butkowski et al. (1977)
β-Thromboglobulin	69	3	97	Begg et al. (1978)
Dihydrofolate reductase	60	7	Not given	Stone et al. (1977)
Dihydrofolate reductase	56	8	96	Bennett et al. (1978)
lac Repressor	54	4	96	Beyreuther et al. (1975)
Azotoflavin	53	6	95.4	Tanaka et al. (1977)
Phospholipase A_2 (peptide)	48	6	Not given	Evenberg et al. (1977)
Histone $H_2B_{(1)}$	47	7	Not given	Strickland et al. (1977)
Collagen peptide	45	9	95	Butler et al. (1977)
Phospholipase A_2	41	7	Not given	Evenberg et al. (1977)
Bungarotoxin	39	2	96.6	Kondo et al. (1978)
Flagellin	39	Not given	97	DeLange et al. (1976)
Uteroglobin	36	5	98	Ponstingl et al. (1978)
Rabbit antibody L-chain	35	Not given	95–98	van Hoegaerden and Strosberg (1978)
Histone H_2B	34	10	95	Kootstra and Bailey (1978)
Pea lectin (α-subunit)	30	6	93	Richardson et al. (1978)
Prethrombin	29	Not given	98	Butkowski et al. (1977)
Ribosomal protein L19	28	1	Not given	Brosius and Arfsten (1978)
Biotin carrier-protein	24	1	Not given	Sutton et al. (1977)

machine is now commercially available and is in use in almost all protein chemistry laboratories throughout the world. The story of Edman's dedication and vision is a fascinating one and has recently been portrayed by one of his close associates (Niall 1977).

Table XIII gives several illustrative instances of the application of commercial sequencers to particular proteins, excluding the most widely used machine, the Beckman 890C sequencer, selected examples from which, taken from the recent literature, are contained in Table XIV. The basic design of all these machines is very similar: they are referred to as 'liquid-phase', or, 'spinning-cup' sequencers to distinguish them from the other available sequencer, the solid-phase modification. As the name 'spinning-cup' implies, these machines are constructed around a central rotating glass cup in which the protein is evenly spread around the walls by the centrifugal force of rotation. The principle is that the thin film of protein so induced provides a large surface area for reactions to occur in high yield—an essential criterion for success in any controlled, repetitive chemical degradation. A simple calculation indicates that the film formed is about 100 molecules thick, and it is this thin molecular layer that facilitates the reaction to occur in high yield.

It can reasonably be assumed that a sequence determination ceases to provide useful information when the over-all yield has fallen to 30%, as at this level the amino acid phenylthiohydantoins cannot be identified against the background, which is partially due to non-specific cleavages of the peptide chain and to incomplete reaction at each step. An indication of this latter factor is conveyed by the term the 'repetitive yield' which theoretically should be 100%, but never attains this in practice. The number of steps possible for any given repetitive yield can readily be calculated. Table XV illustrates the number of steps possible for repetitive yields of 99%, 95% and 90%.

Thus it would appear from this table that with a repetitive yield of 99% at least 100 residues could be sequenced, whereas with a yield of 90% this is drastically reduced to little more than 10 steps. Edman originally reported a repetitive yield of slightly greater than 98% and subsequent workers have little improved on this, although as can be seen from Table *XIV* Frank *et al.*

Table XV. Percentage over-all yield after n steps in the sequential degradation of a protein assuming different repetitive yields

	Repetitive yield		
n	99%	95%	90%
10	90	60	35
50	60	8	0.5
100	36	0.6	0.003

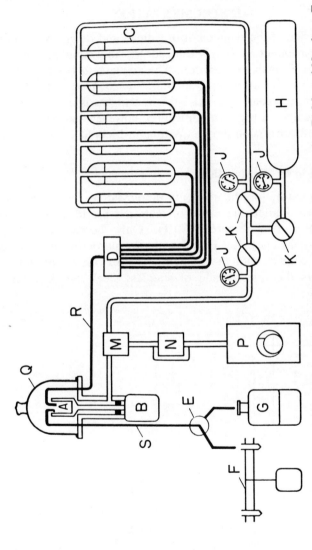

Figure XVII Design of the original sequenator. (A, spinning cup, B, motor; C, reservoirs; D, M, and N valves; E, outlet; F, fraction collector; P, pump; Q, bell jar) (Reproduced from Edman and Begg (1967), by permission of *European Journal of Biochemistry*)

(1978) must have achieved yields of 99% to accomplish a sequence of 100 residues using the Beckman 890C sequencer, but few other workers have attained this degree of success. On average, using the Beckman instrument, most operators obtain a repetitive yield of about 96% and sequence in the order of 40 residues.

INSTRUMENTATION

The general design of the sequenator of Edman and Begg is shown in Figure XVII. This instrument is only designed for the coupling and cleavage reactions of the Edman degradation. The conversion reaction, that is the transformation of the unstable 2-anilino-5-thiazolinone derivatives to the isomeric and more stable 3-phenyl-2-thiohydantoins is performed manually outside the machine. A recent modification, however, has been the addition of an automated conversion device (Wittmann-Liebold 1973).

Materials Used in its Construction

The corrosive nature of the reagents employed in the Edman degradation has severely limited the range of materials useful in the construction of a sequenator (see Figure XVIII). Only borosilicate glass, PTFE (polytetrafluoroethylene). Kel-F (polytrifluorochloroethylene) and gold are used in direct contact with the reagents. Where contact is only with the reagent vapours the finest grade of stainless steel can be used. As the

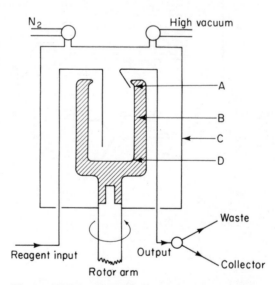

Figure XVIII Simplified view of the reaction cup of the sequenator. A, undercut; B, precision-made glass spinning cup; C, sealed and thermostatically maintained reaction chamber; D, rounded corners

phenylthiocarbamyl group can easily be desulphurized by oxygen, bringing the degradation to an abrupt halt, the reaction is performed in an inert atmosphere, hence the surface area within the reaction cell of the synthetic polymers PTFE, Kel-F etc., and of the delivery system is kept to a minimum as these materials are slightly permeable to oxygen.

The highly inflammable nature of some of the reagents has also meant that all sparking electrical contacts must be excluded, or hermetically sealed: failure to observe these warnings, originally made by Edman, led to disastrous consequences for some of the early versions of the machine.

DESIGN FEATURES

Delivery System

In the original design the delivery system was based on a positive pressure of nitrogen—the reagents being pressurized from a nitrogen cylinder with the

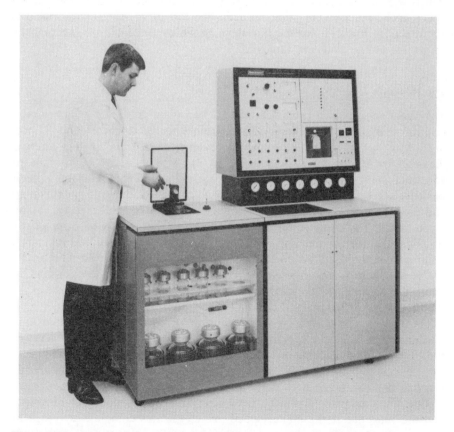

Figure XIX   The Beckman Sequencer 890C. (By courtesy of Beckman Instruments Inc.)

reaction cell kept at a lower pressure. This differential pressure is maintained, thus controlling the volume of reagent or solvent admitted to the cup by the time the appropriate valve is kept open. This principle is also employed in the Beckman sequencer, whereas the JEOL machine uses pumps. The reagents are kept in individual reservoirs each having four connections, one for the nitrogen supply, another for the exit of the chemical, a third for a gas vent and the fourth for refilling. The outlet lines from the reagent and solvent reservoir go to a delivery valve, of which there are four, one for each reagent–solvent pair. Each double port valve and delivery line may be used to deliver a solvent and a reagent, e.g. phenylisothiocynate and benzene; Quadrol and ethylacetate; hepta-fluorobutyric acid and butyl chloride. In this way the solvent washes out the reagent, thus ensuring that no lines become blocked, and as there is no general manifold there is virtually no chance of cross-contamination. All plastic delivery lines are kept as short as possible to minimize oxygen diffusion. In the Beckman machine there is a 'space reducer' (cf Figure XIX) that dips (on a hinged joint) into the reaction cup. This core through which the four reagent–solvent delivery lines are fed, reduces the vapour space in the reaction cup, thus reducing the loss of volatile solvents used in the reaction. It also holds the line through which reagents, solvents and the thiazoline amino acids are removed.

The Reaction Cup

The reaction cup is a precision-made Pyrex-glass container which in the original design had the dimensions 26 mm internal diameter and 31 mm high, however this has subsequently been enlarged to 51 mm internal diameter by 60 mm high (Begg et al. 1978). Attempts have been made to construct cups of PTFE, or other plastics, which have so far only been partially successful, for they have the disadvantage of difficulty in cleaning, and prevent the operator from having a clear view of the reactions taking place inside. The cup is carefully constructed so that the corners are gently rounded to facilitate cleaning. The surface is highly polished and there is a millimetre scale inscribed on the wall to assist calibration. Near the top of the vessel there is an undercut to assist in the scooping operation and to make the accuracy of the delivery of reagents and solvents less critical, as the film height is maintained constant throughout the various cycles.

The cup can be driven by a magnetic drive system as in the Beckman machine, or it might be mounted directly on the end of the motor shaft. In the Edman–Begg machine only one speed was possible (1400 rpm) whereas the Beckman instrument has a variable speed (1200 rpm and 1800 rpm); this has the advantage that as the cup spins faster the level of liquid rises higher. Thus the protein is introduced when the cup is spinning at 1200 rpm and the reagents at the higher speed, which ensures complete coverage of the protein film. The higher speeds are also useful in peptide degradations as the thinner solvent layer formed reduces the losses of slightly soluble peptides. Recently a larger cup, working at a faster speed (3000 rpm) has been introduced (Begg *et al.* 1978).

It is recommended that approximately every 3 months the cup is cleaned in *aqua regia* followed by repolishing the cup surface with cerium oxide.

Drive Mechanism

In the original design the cup was mounted directly on an extended stainless steel motor shaft. However, as it was subsequently found that the seals around where the shaft entered the reaction chamber developed leaks and required regular maintenance, later models of the sequencer employed a magnetic drive assembly. The use of volatile buffers in the sequencing of peptides tends to cause an accumulation of gummy deposits around the drive area and this may cause it to seize up.

Heating System

Two types of heating system have been employed. In the Edman–Begg design the bell jar and the reaction cup were both heated by means of a transparent metallic electroconductive heating layer on the glass surface, whereas in the Beckman machine the glass reaction chamber containing the reaction cup is completely enclosed within a second outer Perspex enclosure, the temperature of which is maintained by circulating hot air from a heater. In some recent models the cap of the reaction cup is heated to a slightly higher temperature than the cup itself, so reducing condensation.

Fraction Collector

The thiazolinones are collected in a refrigerated fraction collector maintained at about 2°C, and in the Beckman machine the contents of the tubes can be taken to dryness by a jet of nitrogen and the compartment automatically evacuated to remove the resulting toxic and inflammable vapours.

REAGENTS AND SOLVENTS

Much emphasis has been laid on the purity of the reagents and solvents. It is vital to remove all traces of aldehydes that might block the sequential degradation by reacting with the terminal amino group. A summary of the purification steps, recommended by Edman and Begg, for the essential reagents and solvents is given in Table XVI. Batches of chemicals that have been pretested on the sequencer may be purchased from several manufacturers. These reagents are extremely expensive and may amount to approximately 10% of the purchase price of the machine per annum.

Table XVI. Recommended purification and storage procedures for reagents and solvents used in liquid phase sequencing

Reagent/Solvent	Purification	Storage
PITC (phenyl isothiocyanate)	Distillation *in vacuo* (1 mm Hg pressure); the fraction distilling at 55° is collected.	In general it is unstable on storage, therefore best to distil as required. It may be stored however, out of direct light, in an evacuated desiccator over P_2O_5.
'Quadrol' (N,N,N',N'tetrakis-2-hydroxypropyl-ethylene diamine)	Purified by passing through a column of amberlite IRA-500 (sulphite) followed by distillation at 120° (10^{-3} mm Hg).	The Quadrol-trifluoroacetic acid salt may be stored for up to a year in amber bottles in a freezer.
n-HFBA (n-heptafluorobutyric acid) TFA (trifluoroacetic acid)	First they are exhaustively oxidized with CrO_3 and then distilled and dried over anhydrous $CaSO_4$. Redistillation is then performed.	Stored in amber bottles in a cold room for up to 6 months. The H_2O content is maintained at 0.01% by careful addition of trifluoroacetic anhydride (Begg *et al.* 1978).
Benzene	Analytical grade is further purified by fractional freezing followed by a distillation (b.pt. 80–81°).	Stored in amber bottles in cold room for up to 6 months.
Butyl chloride (1-chlorobutane)	Purified by passage through a column of activated granular charcoal followed by distillation (b.pt. 78°).	Stored in amber bottles in cold room for up to 6 months.
Ethyl acetate	Passed through a column of activated Al_2O_3, redistilled and dried on a molecular sieve (Type 4A, Union Carbide).	Stored under nitrogen (<10 ppm of O_2) in amber bottles in cold room for up to 6 months.
Heptane	Treated overnight with an aqueous solution of $KMnO_4$, washed with H_2O, then dried over Na_2SO_4 and distilled.	Stored in cold room in amber bottles for up to 6 months.

PROCEDURES

The sequence analysis of large proteins and smaller peptides requires different tactics. Proteins are sequenced in non-volatile buffers and reagents, whereas peptides require volatile reagents. Each of the approaches will now be considered in detail.

(1) The Automatic Degradation of Proteins

Sample Preparation

The freeze-dried protein sample must be salt-free and may be in its native state, however, better results are obtained if it is completely denatured either by performic acid oxidation, or following reduction and alkylation of the cysteine residues. It is best if the protein is soluble in the reagents used, though successful sequences have been obtained on proteins insoluble in the reagents.

Programmes

There are two possible programmes that can be used when sequencing proteins: the single cleavage and the double cleavage programme. Essentially, the double cleavage programme, which was used successfully by Kondo *et al.* (1978) in their study of the sequence of the A and B chains of β-bungarotoxin, puts the protein through the acid cleavage reaction twice. This ensures complete removal of the N-terminal residue, but only one extraction with butyl chloride is collected as there is insufficient material in the second extraction to justify its collection, so this is diverted to waste. An outline of the seven steps in the double cleavage programme is contained in Figure XX.

Reagent Modifications

(i) THEED NNN'N' Tetrakis (2-hydroxyethyl)ethylene diamine (THEED) has been used in place of Quadrol in the coupling reaction (Begg and Morgan 1976). It has the advantage that it may readily be prepared free of contaminating aldehydes that tend to reduce the repetitive yield when Quadrol buffers are employed, for example regular repetitive yields of 98% are obtained in contrast to the 94% when quadrol buffers are used. However there are problems associated with the use of THEED; impurities present in it may be extracted by the chlorobutane into the thiazolinone fraction and these interfere with the TLC identification of the PTH-amino acids. It is therefore suggested that when this buffer is employed, back hydrolysis of the PTH be used for the identification of the amino acid residue.

The preparation of THEED starting from ethylenediamine and ethylene oxide has been described (Begg *et al.* 1978).

58

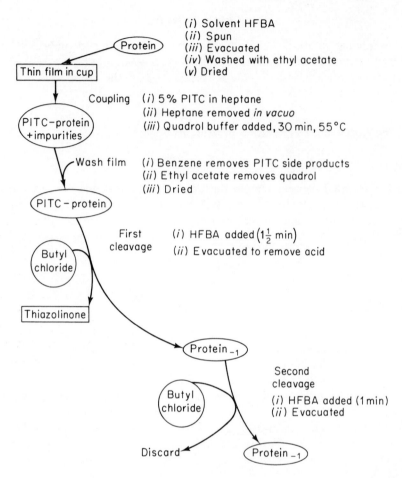

Figure XX   An outline of the 'double cleavage' programme in liquid phase sequencing

(ii) Pentafluoropropionic Acid   Inglis and Burley (1977) in their study of the sequence of Apovitellenin I from duck eggs found that hepta-fluorobutyric acid was not readily removed after the cleavage reaction and consequently the residual acid caused a loss of hydrophobic peptides during thiazolinone extraction. This problem was overcome by using the more volatile pentafluoropropionic acid for cleavage and the mixture 1,2 dichloroethane–benzene (3 : 1) to extract the thiazolinone.

Programme Modifications

(i) Preconditioning   Kootstra and Bailey (1978) found that improved yields could be obtained if the protein film in the spinning cup was subjected

to one complete degradation cycle in the absence of PITC to precondition the protein film.

(ii) Cleavage of Prolyl Residues   Where prolyl residues are known to occur from preliminary experiments the cleavage time is extended threefold to avoid incomplete cleavage (Begg *et al.* 1978).

(iii) Addition of PITC before the Coupling Buffer   Thomsen *et al.* (1976) report that non-specific cleavage can be reduced if the usual order of addition of PITC and coupling buffer is reversed. They claim that these cleavages are due to acid-catalysed N → O acyl shifts involving seryl hydroxyl groups.

(iv) The Repetitive Yield and Length of Run   The repetitive yield obtained by Edman and Begg was slightly greater than 98% and allowed 60 residues of apomyoglobin to be determined. This performance has rarely been equalled by the many commercial instruments available (Table XIII). The reasons for this must be a combination of operator inefficiency and the use of inferior quality reagents.

That some of the commercial machines are capable of a greater efficiency has been shown by Frank *et al.* (1978), who sequenced 100 residues of a protein with the Beckman 890C sequencer. The repetitive yield in this case must have been almost 99%. Most operators, however, achieve yields of about 95% that make possible a run of about 40 residues; a typical example is shown in Figure XXI.

Sometimes even these sort of yields are not achieved, resulting in a gradual carry-over of amino acids that increase during the run, for example in the sequence determination of S-pyridylethyl (insulin-like) Growth Factor (Rinderknecht and Humbel 1978) the carry over was 14% at step 3 rising to 72% at step 25; however, this amount of carry-over did not prevent

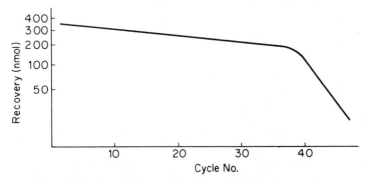

Figure   XXI   Automated   Edman   degradation   of   rabbit uteroglobin, showing the gradual decline in yield. (Adapted from Ponstingl *et al.* 1978)

identification up to residue 31. Nevertheless, better results ought to be possible, and in those few cases where the protein presents particular problems, duplication of the sequence run is essential ('Documentation of Results in The Determination of the Covalent Structure of Proteins'. *J. Biol. Chem.* (1976) **251**, 11–12).

(2) The Automatic Degradation of Peptides

The procedures used for the automatic sequence analysis of proteins cannot be applied to peptides. Peptides, particularly when they are hydrophobic in nature, tend to be lost in the extraction into the organic solvents. Several attempts have been made to modify the procedures to avoid these losses. Two factors must be taken into account: first, as a peptide is smaller than a protein there are therefore fewer non-specific cleavage points which results in a clearer background than that found for proteins; secondly, as there will be fewer polar groups present than in a protein, and taking into account its smaller size, the peptide will be lost to a greater extent during the extraction process into the organic solvent, than for a protein. The programme employed therefore should reduce the amount of solvent washes and, in the main, should involve only a single cleavage. It has been found that most of the losses occur during the extraction with butyl chloride following the acid cleavage, and to a lesser extent during the extraction with ethyl acetate. The peptides most likely to be lost by this 'washing out' will be those hydrophobic ones and those having several lysine residues, the ε-amino groups of which react with the PITC to form phenyl thiocarbamoyl derivatives, so becoming more soluble in the organic solvent. The following attempts have been made to prevent this 'washing out'.

(1) The Use of Volatile Buffers

The replacement of quadrol by a volatile buffer would remove the necessity of an ethyl acetate wash. Two such buffers have been used: NN-dimethylbenzylamine (DMBA) and NN-dimethylalkylamine (DMAA). Of these the DMAA buffer is the more volatile and causes problems in the coupling, as it tends to evaporate, leading to a reduction in the pH of the reaction medium. Deposits have also been found to form around the drive mechanism, which necessitates frequent cleaning of this area, and the oil in the vacuum pump requires frequent replacement when this buffer is used: however, it has been successfully used in a number of investigations. Macleod *et al.* (1977), in their study of trout-testis histone H1, employed DMAA buffer and found that improved repetitive yields could be obtained; Stone *et al.* (1977) also used this buffer for the sequencing of peptides from dihydrofolate reductase. The less volatile DMBA has also found considerable application, amongst others, in studies of the peptides from the neurotoxins of the African cobra (Gregoire and Rochat 1977): this work was achieved with a SOCOSI sequencer. Rochat *et al.* (1976) found that DMBA enabled as little as 20 nmol of a pentapeptide to be completely sequenced.

(2) Modification of Lysine Residues

(i) Reaction with 4-sulpho-phenylisothiocyanate converts the ε-amino group of lysine to a sulphophenylthiocarbamoyl derivative imparting

$$N=C=S$$

$$SO_3^- \ Na^+$$

improved hydrophobicity to the peptide chain. With this modification it is possible to obtain high degradation yields using the regular PITC-quadrol buffer; this was the technique used by Mak and Jones (1976) in their study of wheat β-purothionin, and Stone *et al.* (1977) in studies on dihydrofolate reductase. Strickland *et al.* (1977) found that if peptides are 'anchored' into the cup by reacting the ε-amino groups with 4-sulpho-phenylisothiocyanate, yields of up to 90.3% were obtained making it possible to sequence up to 37 residues of a cyanogen bromide peptide isolated from the histone $H_2B_{(1)}$ from sea urchin sperm. van den Berg *et al.* (1976) used this reagent to sequence peptides from coypu RNase; the peptides after modification were desalted on a column of Sephadex G15 and then sequenced in a JEOL sequencer using the 'S4 programme' in which 10% of the residual peptide is withdrawn from the cup and the N-terminal identified by dansylation.

Dwulet and Gurd (1976) have recommended the use of the *meta*-isomer of SPITC viz. 3-sulpho-phenylisothiocyanate, which has several advantages over the 4-isomer, including a greater coupling efficiency, the repetitive yield increasing by 2–3%, and the fact that the sequencing product, namely ε(3-SPTC)-PTH-lysine can be converted to free lysine after back-hydrolysis more readily than from the derivative of 4-SPITC.

$$N=C=S$$

$$SO_3^- \ Na^+$$

(ii) Braunitzer Reagents Braunitzer and co-workers have introduced a number of improved derivatives to modify the ε-amino groups of lysyl residues (as shown in the structures) (Braunitzer *et al.* 1971, 1973).

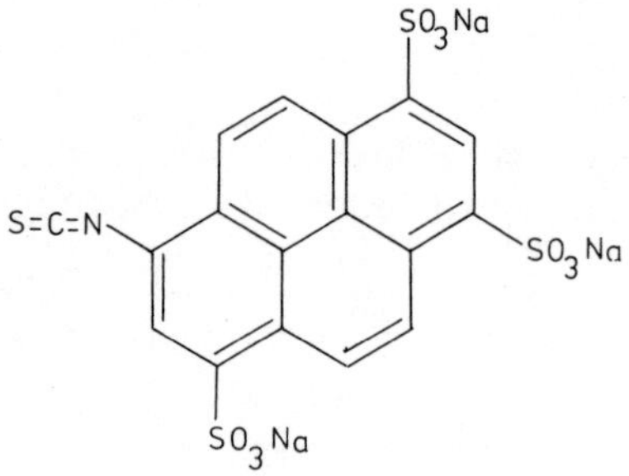

(i)

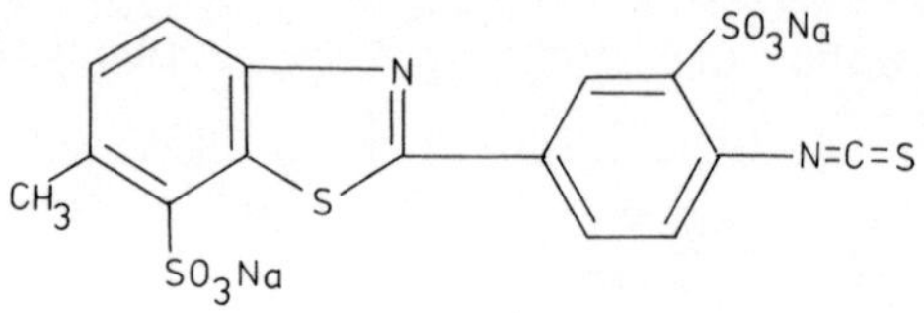

(ii)

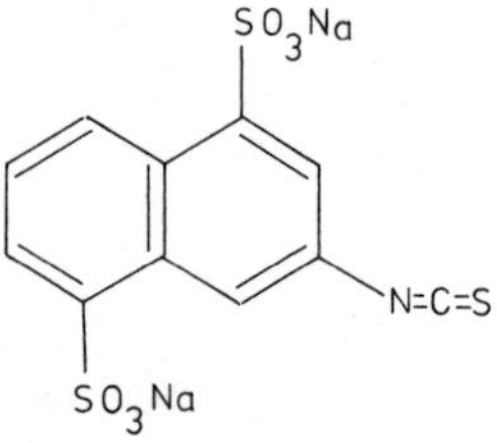

(iii)

Reagent (iii) permits high degradation yields with the normal quadrol buffer system, and the peptides derivatized have excellent film-forming characteristics that are important in automated sequencing.

(3) Modification of Acidic Peptides

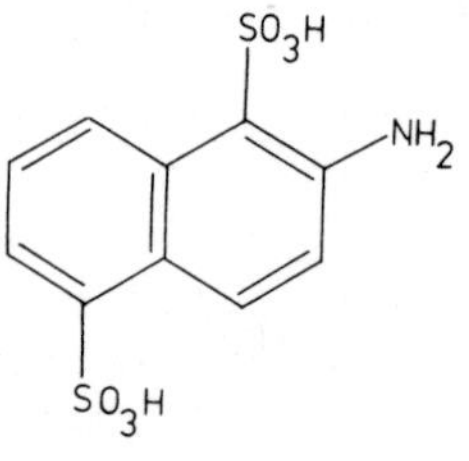

Acidic or neutral peptides lacking lysyl residues can be modified by reaction of the carboxyl groups with the reagent 2-amino-naphthalene 1,5-disulphonic acid. This renders the peptide more hydrophilic and less likely to be lost during extraction into the organic solvents during the sequencing. This reagent is coupled to the peptide by using a water-soluble carbodiimide (Foster *et al.* 1973), the most usual being N-ethyl, N'-(3-dimethyl aminopropyl) carbodiimide (Dixit *et al.* 1977a,b). (Care

$$CH_3CH_2-N=C=N-CH_2CH_2CH_2N\begin{array}{c} \diagup CH_3 \\ \diagdown CH_3 \end{array}$$

must be taken when handling these reagents, in particular they must not be allowed to come into contact with the skin or eyes). As with the SPITC reagents the modified amino acid residues cannot be identified as PTH derivatives during the automatic sequencing and therefore have to be identified by back hydrolysis.

(4) Modification of Cysteine-containing Peptides

Peptides containing cysteine residues may be modified by reaction with the sodium salt of 2-bromoethane sulphonic acid prior to sequencing; however, as cysteine is not a very common amino acid residue in proteins it is not as useful as other available methods (Niketic *et al.* 1974). Alternatively, cysteine may be converted to the polar cysteic acid after performic acid oxidation, but this has the disadvantage of destroying any tryptophan residues that might be present.

Modified Programmes for Use with Peptides

Hogg and Hermodson (1977) introduced a slightly modified 'peptide programme' that enabled the sequence of peptides to be determined to within several residues of the C-terminus. In principle this involved the introduction of steps designed to precipitate the peptide with the organic solvents. For example, benzene was allowed to fill the reaction cup and remain in it for an interval of 20 s, so as to complete the precipitation, and this was subsequently followed by a precipitation step with chlorobutane. Similar tactics were employed by O'Donnell and Inglis (1974) in their study of feather keratin, that enabled them to sequence 53 residues of a large peptide. They also found it advantageous to employ a prolonged drying period following the cleavage with acid; this was found to reduce the 'washing-out' of the peptide during the extraction of the thiazolinone.

An alternative strategy was used by Tanaka *et al.* (1975) in their study of *Spinulina maxima* ferredoxin. In this investigation, the repetitive yield by step 29 had so dramatically dropped that they imaginatively withdrew the peptide from the cup and performed a further 18 additional steps manually. This was the first time this had been done. Furthermore subsequent workers have combined the conventional dansyl method, and the automatic Edman procedure by taking 10% of the peptide material from the cup at each stage for N-terminal analysis. (van Den Berg *et al.*

1977). It would appear therefore that in the future automated and manual methods might be complementary to each other.

THE USE OF CARRIERS

The amount of peptide needed to form a stable film in the cup is between 1 and 2 mg. Thus for a peptide of 50 residues, about 200 nmol is required, and as it is essential to repeat the sequencing at least once then a total of about 400 nmol is needed, i.e. about 2 mg.

For smaller peptides this amount is far in excess of that needed for manual determination, so to overcome this problem the use of an inert 'carrier' has been introduced. The ideal properties of a carrier should be:

(1) stability, during the sequential degradation;
(2) polarity, so as to reduce loss during the solvent wash;
(3) ability to form a stable thin film that is both firm and even.

The two types of carrier that have so far been used are either synthetic polymers, or small proteins with blocked amino termini. Of this latter type, the French group at Marseille have employed hake parvalbumin in their SOCOSI sequencer with considerable success. Martinez *et al.* (1977) have sequenced up to residue 26 of a polypeptide toxin in the presence of 2 mg of parvalbumin. This technique made it possible to sequence as little as 20 nmol of peptide, thus enhancing the sensitivity of the automated method (Rochat *et al.* 1976).

The Use of Synthetic Polymers

(1) Polybrene

'Polybrene' is the trade name for the polymer 1,5-dimethyl-1,5-diazaundecamethylene polymethobromide. When as little as 3 mg of this polymer is added directly to the peptide solution it has been found to effectively retain small peptides in the spinning cup (Tarr *et al.* 1978). This has the advantage that no chemical modification of the peptide is necessary and enables a high repetitive yield to be obtained. Using this innovation Frank *et al.* (1978), were able to sequence small peptides right up to the C-terminal residue, in their study of the structure of C-Phycocyanin.

(2) Succinylated Poly-ornithine

The successful use of succinylated poly-L-ornithine as a 'carrier' in the automatic sequencer has been described (Silver and Hood 1974). The addition of this synthetic material to the peptide reduces the 'washing-out' of the peptide material and assists in the holding of small amounts of hydrophobic peptides in the cup. Inglis and Burley (1977), in their study of the sequence of egg apovitellenin, added 5 mg of this carrier to 500 nmol of a CNBr peptide and it enabled them to completely determine its sequence.

(3) (Norleu–Arg)$_{27}$

Niall *et al.* (1974) have suggested the use of the synthetic polymer (Norleu–Arg)$_{27}$ as a carrier. Using this substance, a linear rather than an exponential decrease in the yield was found with myoglobin as a trial protein. One problem, however, was the release of arginine from the carrier, which caused confusion when arginine was present in the sequential degradation. Attempts have been made to produce a carrier based on homoarginine.

(4) Chemically Modified Lysozyme

Frank *et al* (1978) have added a modified lysozyme (lysozyme reacted with sulphobenzoic acid anhydride to block its N-terminus) to approximately 300 nmol of peptide derived from C-Phycocyanin. It was found to successfully hold the peptide in the spinning cup during the sequential degradation.

CHAPTER 5

The Solid-phase Sequencer

The solid-phase sequencer was designed primarily for the automatic sequencing of peptides containing approximately 30 amino acids or less, i.e. those that would tend to be 'washed-out' in the spinning-cup sequenator. Laursen (1971) was the first person to design such a machine and his publication gave extensive details that have enabled other groups to construct their own machines (Wiman 1977). Also there are at least three commercially available versions of this instrument (Table XVII), and furthermore it is also possible to convert the liquid-phase sequenator into a solid-phase machine, this being easier to accomplish with the Jeol sequenator; nevertheless a permanent conversion would not be sensible from an economic point of view. This is because the principal attraction of the solid-phase sequencer is its low cost compared to the spinning-cup instrument, and this low capital cost reflects the overall simplicity of this machine. Although simple in construction, the solid-phase sequencer has not had the overall success the liquid-phase model has achieved. Possibly this reflects the smaller number of these instruments in operation, but more realistically it is a consequence of the involved chemical manipulations that are necessary in this method of automatic sequencing. Although not completely satisfactory in themselves, more development being needed in this area, the additional chemistry necessary has not attracted workers to adopt this method of sequencing. However, the relatively low cost of the solid-phase sequencer has in more recent years attracted interest, particularly from the individual worker who could not justify the massive capital cost of a spinning-cup sequencer.

It is clear that if the chemical steps necessary for solid-phase sequencing

Table XVII. Solid-phase sequencers and their manufacturers

Instrument	Manufacturer
SEQUEMAT (Models 10K and 12)	Sequemat Inc., Massachusetts, USA.
LKB (Model 4020) (see Figure XXII)	LKB Biochrom. Ltd, Cambridge Science Park, U.K.
SOCOSI (Model PS300)*	SOCOSI, France.

*Also sold in kit-form for self-assembly.

could be simplified, and improved in effectiveness, there could be a revival of interest in this method of sequencing.

GENERAL PRINCIPLE

The peptide must be covalently bound to an insoluble residue to be available for reaction with the Edman reagent. This and other reagents are pumped over the immobilized peptide and the appropriate fraction containing the anilinothiazolinone collected for identification. 'Washing-out' of the peptide by the reagents is thus completely avoided. This elegant method of sequencing was originally seen in primitive form in the 'paper-strip' procedure devised by Fraenkel-Conrat *et al.* (1955), which although a brilliant innovation, that was designed in an attempt to meet the expanding demand for sequence information, never 'caught-on' universally due to several minor difficulties.

INSTRUMENT DESIGN

The basic design of the solid-phase sequencer is shown in the simplified block diagram and the more detailed design of the original version of Laursen's is shown in Figure XXIII.

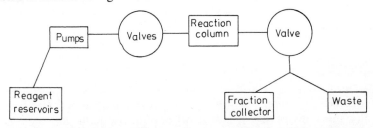

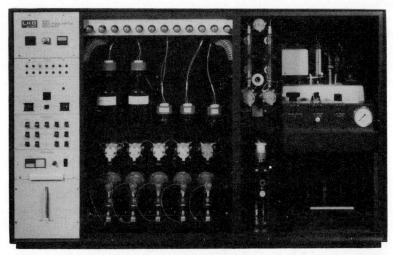

Figure XXII The LKB solid phase sequencer (by courtesy of LKB Biochrom Ltd, Cambridge)

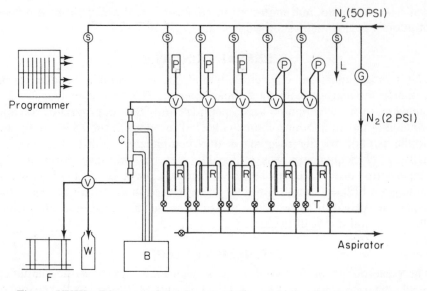

Figure XXIII Diagram of the layout of the original solid phase sequencer. (B, water bath; C, column; F, fraction collector; G, gas cylinder; P, pump; R, reservoir; V, S, valves.) (Reproduced from Laursen (1972), by permission of Academic Press, Inc.)

Reaction Column

The reaction chamber in the LKB and Sequemat instruments is a jacketed glass microbore column in which is situated the immobilized peptide. Normally the bulk of the resin-bound peptide is small and it is usual to enhance this by the addition of inert glass beads; this also, in the case of polystyrene-based supports, reduces the effects of the resin swelling and shrinking in the reaction column. Two columns are usually employed, and in some cases these have different internal diameters that permit the accommodation of distinct supporting materials. To maintain the column at a constant temperature, which is normally 45°C, water is circulated through the surrounding jacket by an integral water bath.

The Shaking Reaction Chamber

The reaction chamber in the French SOCOSI (PS300) machine, instead of being a column, consists of a unique shaking chamber. This vessel, which is thermostatted at 45°C, is shaken vertically. It has an internal volume of approximately 3 ml and to ensure efficient shaking the reagent volume should be more than 1.5–2 ml. This clever design eliminates the necessity for high-pressure pumps that are needed in the column version, the reagents being stored and pushed through by pressurized nitrogen gas. The disadvantage of this design, apart from the low washing efficiency achieved,

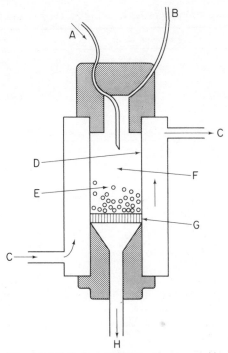

Figure XXIV Simplified drawing of the shaking reaction vessel of the SOCOSI sequencer. (A, inlet reagent tube; B decompression tube; C, circulating water bath; D, internal facings coated with Kel-F; E, resin particles; F, reaction chamber; G. sintered glass filter; H outlet)

is that reagents may become entrapped in the joints and glass filter of the reaction vessel.

MATERIALS AND REAGENTS

The materials used in the construction of the sequencer, like the spinning-cup sequenator, are governed by the high chemical reactivity of the reagents of the Edman degradation, therefore all components that come into direct contact with these reagents are constructed of Teflon, Tefzel, Delrin, Kel-F, glass or stainless steel. The pumps in the LKB version are made entirely of glass and PTFE, with pistons of sapphire.

The demand on reagent purity is not as great as that necessary with the spinning-cup machine, by virtue of the fact that impurities cannot accumulate but are continuously washed through. In fact Laursen (1971) found ordinary reagent grade materials satisfactory; nevertheless for extended sequencing it is desirable to have analytical grade reagents. Even

Table XVIII. Preparation and properties of supports used in solid-phase sequencing

Support	Preparation
Amino polystyrene	Polystyrene (400 mesh) —HNO₃ (95%), 1 h, 0°C→ Nitropolystyrene (·NO₂) —SnCl₂/DMF, 140°, 15 min→ Aminopolystyrene (NH₂)
Triethylene tetramine polystyrene	Chloromethylpolystyrene (—CH₂Cl) —NH₂(CH₂)₂NH(CH₂)₂NH(CH₂)₂NH₂, 100°C, 90 min→ Triethylene tetraamine-polystyrene (—CH₂—NH(CH₂)₂—]₃NH₂)
3-Amino propylglass	Pourous glass (400 mesh, 75 Å pore diam.) (—OH, —OH, —OH) + EtO—Si(OEt)(OEt)—(CH₂)₃NH₂ (3-Amino propyl triethoxysilane in dry acetone) —25°→ 3-Amino propyl glass
N(2-aminoethyl)-3-aminopropylglass	Porous glass (400 mesh, 75 Å pore diam.) (—OH, —OH, —OH) + MeO—Si(OMe)(OMe)—(CH₂)₃NH(CH₂)₂NH₂ (N-(2-aminoethyl) 3-amino propyl trimethoxy silane (4% in acetone)) —25°→ (—O, —O, —O)Si(CH₂)₃NH(CH₂)₂NH₂

this entails a considerable saving on the expense of the high-purity reagents needed in the liquid-phase machine.

Types of Support

Much thought and ingenuity has been given in the search for an ideal type of support. Up to date no such material has yet been discovered and the ones that are available must be employed with discernment. In general, the requirements for a support are threefold:

(1) it must be inert to the wide range of chemicals employed;
(2) it must be stable to the mechanical stresses involved; and
(3) it must be permeable to both peptides and reagents.

A selection of supports that have been successfully employed are listed in Table XVIII, which also gives details of their preparation. The strategical problems involved in choosing a support must include a number of factors, in particular the size of the peptide must be taken into account, as well as the method of chemical attachment to be employed. Polystyrene-based supports, although adequately supplied with amino groups, do not swell in aqueous buffers, which limits their use to those peptide derivatives that are soluble in organic solvents. Whereas porous glass derivatives, on the other hand, cannot be highly substituted with amino groups and the coupling of

Colour	Storage conditions	General comments
Brown	Very stable and may be stored for long periods under refrigeration	Best used when the coupling reagent is p-phenylene diisothiocyanate. Suitable for peptides <35 residues.
White	Unstable and best prepared as needed; however it is stable for several months in deep freeze.	The best choice if coupling the peptide by its c-terminus. It is particularly effective for coupling homoserine lactone activated peptides. Suitable for peptides <70 residues.
Yellow	Stable at 4°C for at least 6 months.	Suitable for large peptides and small proteins, however the coupling procedure employed is less critical than for the polystyrene-based supports. They are not suitable for peptides <15 residues as these tend to absorb on to the glass surface.
Gold	Stable at 4°C for at least 6 months.	

the peptide is generally inefficient. The answer to these problems is a hydrophilic support that would swell in both aqueous and organic media. One such polymeric material is polyacrylamide, however this has the disadvantage that it is degraded in the presence of trifluoroacetic acid. To overcome these problems there has been some recent work on the modification of this polymer, with the introduction of the resins: β-alanyl-hexamethylene diamine polydimethyl acrylamide, and N-aminoethylpolyacrylamide (Atherton *et al.* 1976, Cavadore *et al.* 1976).

The Chemistry of the Coupling Procedures

There have been many attempts to devise a suitable means for the covalent coupling of the peptide, or protein, to the inert support. Much of this has been possible due to developments that have occurred in the field of solid state peptide synthesis. There are basically four approaches possible, for attaching the peptide to the inert support; these are:

(1) attachment through C-terminal homoserine lactone;
(2) attachment by activation of the C-terminal carboxyl group;
(3) attachment through side-chain amino groups; and
(4) attachment of the protein to the support *via* cysteine residues. Each of these will now be considered in detail.

(1) Attachment Through C-terminal Homoserine Lactone

Cyanogen bromide fragments, excepting the one derived from the C-terminal region of the protein, terminate in homoserine. By treatment with anhydrous trifluoroacetic acid the C-terminal homoserine residue is converted quantitatively into the lactone, viz.

terminal homoserine   terminal homoserine lactone

The lactone ring formed is activated sufficiently to undergo rapid aminolysis with a polymeric amine. In a typical experiment the peptide after treatment with TFA is dried down and coupled directly onto a triethylenetetramine derivative of polystyrene by simply stirring in dimethylformamide. Following coupling, excess amino sites on the resin are blocked with methyl isothiocyanate $(CH_3 N = C = S)$ and the peptide sequenced.

peptide with terminal homoserine lactone

coupled peptide

This is a very elegant and simple procedure. The product is formed in high yield and the method has the advantage that during sequencing of the coupled peptide no gaps are to be expected, in contrast to other procedures. However it is only applicable to cyanogen bromide peptides which does limit its usefulness.

(2) Attachment by Activation of the C-terminal Carboxyl Group

All the methods available for the attachment of a peptide to an insoluble support by means of its terminal carboxyl group present the problem that no matter what procedure is used, inevitably side-chain carboxyl groups will also be attached to the resin. One solution to this dilemma is to hydrolyse the protein into peptides that terminate in either glutamic acid or aspartic acid. This may be accomplished by using the staphylococcal, or 'Glu' protease. Another ingenious approach for tryptic peptides has been to react all the available carboxyl groups with glycinamide in the presence of a carbodiimide. The terminal carboxyl group may be specifically unmasked by reacting the peptide with trypsin (see Figure XXV).

The method, however, has the disadvantage that the modified internal residues of glutamic and aspartic acids cannot be directly identified as their PTH derivatives, and indirect methods such as back hydrolysis are adopted.

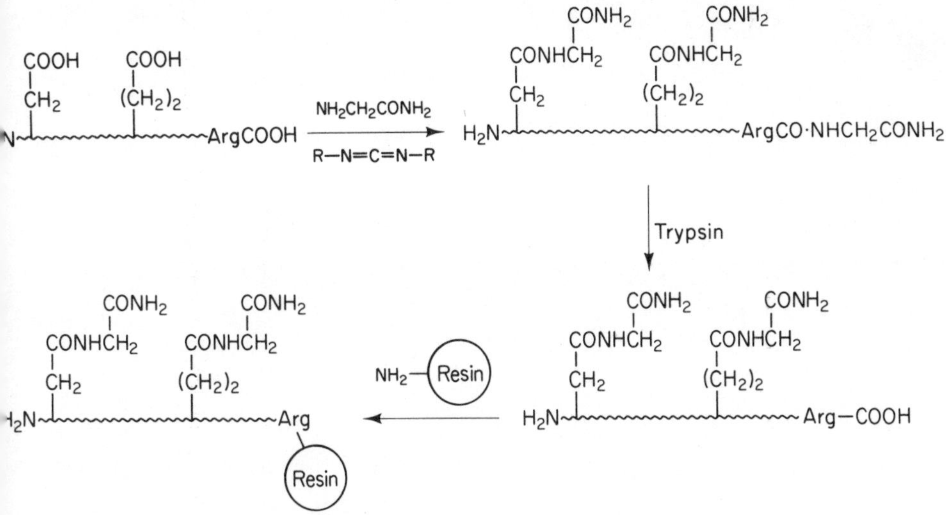

Figure XXV   Blocking side chain carboxyl groups with glycinamide

(i) Using N,N'carbonyldiimidazole   Historically this is an important method for coupling the peptide to the inert support as it was one of the first methods used (Laursen 1971). N,N'carbonyl-diimidazole (mol.wt 162.16, m.pt 116°–118°) reacts readily with carboxylic acids forming imidazolides, that undergo rapid aminolysis with an amino polymer (see Figure XXVI). This method requires both protection of amino side chains in the peptide and the use of anhydrous solvents. The amino group is best protected by t-butyl azidoformate which is an excellent blocking group for it may be readily removed in the presence of trifluoroacetic acid.

Despite coupling efficiencies of greater than 80% (Laursen 1971) this method has received little attention. The use of carbonyldiimidazole

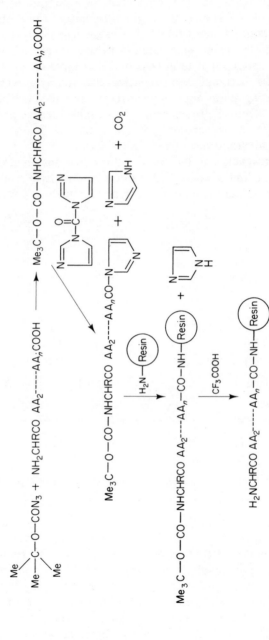

Figure XXVI   Coupling peptides to insoluble membranes using N,N′carbonyl diimidazole

demands stringent anhydrous conditions, which are on the whole unsatisfactory with the majority of water-soluble peptides. A further problem is the additional work involved in the protection of the amino terminal residue. Nevertheless both these problems have been elegantly avoided by Beyreuther et al. (1978), who first subjected the peptide to reaction with phenylisothiocyanate followed by coupling under anhydrous conditions to 3-aminopropylglass. Finally, there is the problem that side chain carboxyls are modified when coupling with carbonyl diimidazole: the glutamyl residue is bound to the support which leads to gaps in the sequence; whereas aspartyl residues undergo cyclic imide formation, thereby bringing the sequencing procedure to a premature end.

(ii) Using carbodiimides Carbodiimides are useful reagents for the coupling of peptides to supporting media, under conditions that favour selective blocking of the side-chain carboxyl groups (Previero et al. 1973). As a preliminary step it is essential to block terminal amino groups with a reversible blocking group, such as t-butoxy-carbonylazide. The success of this method depends on the ability of the investigator to exclude all extraneous nucleophiles during the activation process, thereby reducing the number of side products. Previero et al. (1973) found that under certain conditions the carbodiimide could provide selective blocking of side-chain carboxyl groups.

For this purpose water-soluble carbodiimides are used, e.g. N-ethyl-N'(3-dimethyl aminopropyl)carbodiimide. It was found that this reagent reacts initially with side-chain carboxyls to give an O-acyl urea derivative which subsequently rearranges to give an inactive N-acyl urea, whereas the C-terminal carboxyl group reacts with the carbodiimide to form an oxazolinone which is still reactive to suitable nucleophiles, such as amino groups. In theory this is good, however in practice experience shows that side reactions may occur and the coupling efficiency obtained is quite variable (see Figure XXVII).

This technique was employed by Lindemann and Wittman-Liebold (1977) to couple tryptic peptides from the ribosomal protein S13, to aminopolystyrene and enabled peptides in the range 30–100 nmol to be successfully sequenced. Also, Beyreuther et al. (1977) in their studies of the structure of the phosphocarrier protein HBr of the phosphoenol-pyruvate-dependent phosphotransferase system of *Staphylococcus aureus*, used carbodiimide to couple peptides in the range 20–100 nmol to an inert support. These workers used the novel technique, that was subsequently employed by K. J. Dilley (Protein Chemistry Note 6, LKB Ltd) in sequencing chick histone H_5, of blocking the amino residue with phenylisothiocyanate. The phenylthiocarbamoyl peptide derivative was extracted into benzene and activated by carbodiimide in dimethylformamide before coupling to the support. This method eliminates the need for a subsequent unblocking step as the Edman degradation may proceed once coupling is complete.

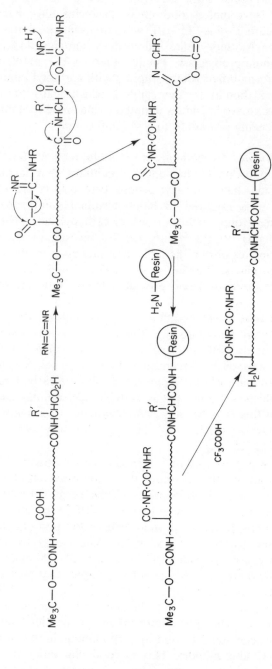

Figure XXVII Coupling peptides to insoluble membrane with carbodiimide

(3) Attachment by Side-chain Amino Groups

(1) Use of p-phenylene diisothiocyanate Peptides and proteins may be bound covalently to aminated supports using the bifunctional reagent p-phenylene diisothiocyanate (DITC). In the coupling process it is essential to use a large excess of reagent so as to prevent cross-linking, or even polymerisation of the peptide. Coupling yields of up to 100% have been reported (Laursen *et al.* 1972); however, this method is limited to those peptides containing lysine, or arginine.

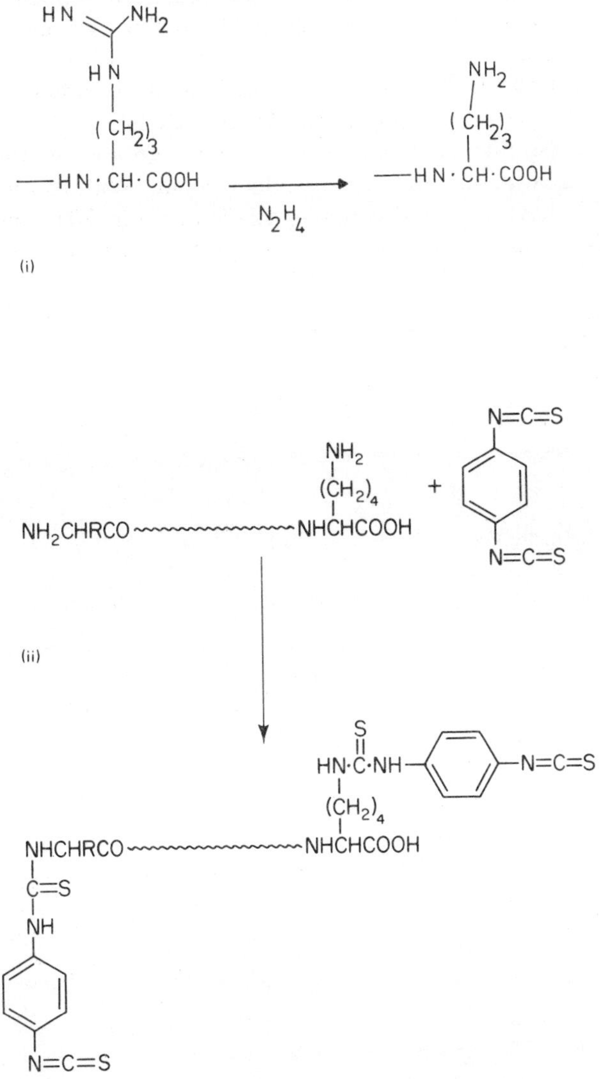

Figure XXVIII (i) Conversion of Arg→Orn. (ii) Activation of peptides with *p*-phenylene diisothiocyanate

The arginine must, prior to coupling, be converted to ornithine. This is usually performed with aqueous hydrazine (50%, at 70° for 15 min): however, the author has found these conditions to occasionally lead to chain cleavage and it is advisable to perform a preliminary small-scale experiment before commiting the whole of the sample to this treatment. Laursen *et al.* (1972) have also reported difficulties with the hydrazine treatment; in particular it is essential to remove all traces of the base before coupling as this will reduce the coupling efficiency.

Coupling is actually performed in two stages; first the peptide is activated by treatment with a 50–100-fold molar excess of p-phenylene diisothiocyanate which reacts with all available amino groups (as shown in Figure XXVIII).

In the second stage, coupling is performed by adding the aminated support to the activated peptide (see Figure XXIX).

Once the peptide is bound, excess amino groups on the support are blocked by treatment with methyl isothiocyanate. With this technique Chen *et al.* (1978) have reported coupling yields of up to 70%, involving tryptic peptides from human myeloma λ-chain protein to aminopolystyrene. One

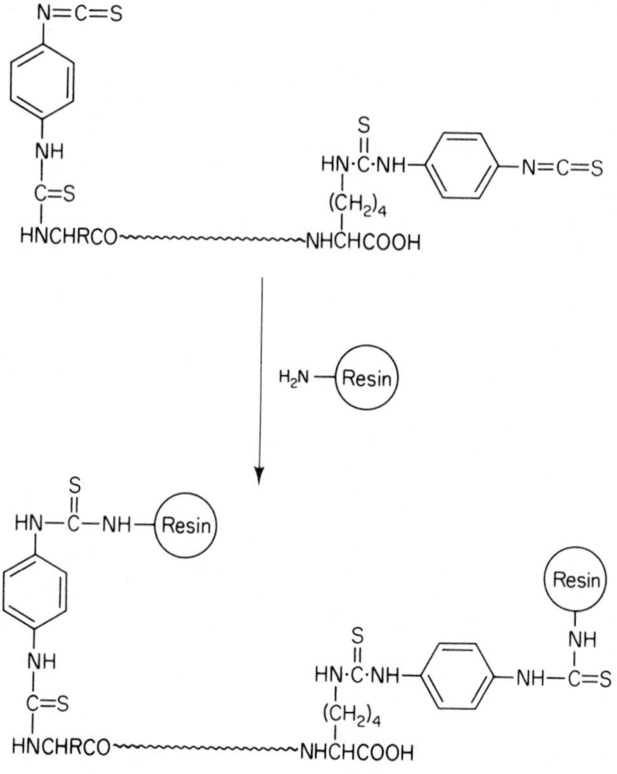

Figure XXIX Coupling of activated peptides to an
aminated support

point that must be noted is that the lysyl/ornithyl, and amino terminal residues, are irrevocably bound to the resin and this results in gaps in the sequence. These drawbacks, however, did not prevent Hitz *et al.* (1977) sequencing the ribosomal protein S6 up to residue 34, after it had been attached to aminopolystyrene. It is interesting to note that, in comparision, they were able to sequence up to residue 44 with a Beckman liquid-phase sequencer, slightly more sample being required in this case.

(ii) Use of N(p-isothiocyanatobenzoyl) DL-*homoserine lactone* Herbrink (1976) introduced the bifunctional reagent N(p-isothiocyanatobenzoyl)

DL-homoserine lactone for the specific attachment of peptides to an aminated support (see Figure XXX). This reagent contained two functional groups differing in reactivity. Using this reagent, cross-linking of peptides, which is occasionally observed with bifunctional reagents bearing two identical reactive groups, is virtually eliminated. This reagent therefore combines the reactivity of the di-isothiocyanate method with the specificity of the homoserine lactone method. Herbrink successfully applied this procedure to a number of tryptic peptides from the β-Bp-chain of bovine β-crystallin.

(4) Attachment of Proteins to Inert Supports

(i) Use of p-phenylene diisothiocyanate A slightly different technique is employed than that with peptides. The support, rather than the peptide, is activated with the reagent (see Figure XXXI)

In a typical experiment p-phenylene diisothiocyanate is dissolved in dimethylformamide to give a 25–100-fold excess of reagent over the available amino groups on the aminated support. The aminopropyl glass is added in small portions under a nitrogen atmosphere over a period of one hour and the product is filtered and carefully washed. The activated support is then added to the protein dissolved in a small volume of aqueous pyridine containing N-methyl morpholine. After 10 min at room temperature the product is filitered off and washed.

(ii) Attachment of proteins to inert supports by their cysteine residues Chang *et al.* (1977) found it possible to attach proteins *via* their cysteine residues to an iodoacetamide derivative of porous glass beads (see

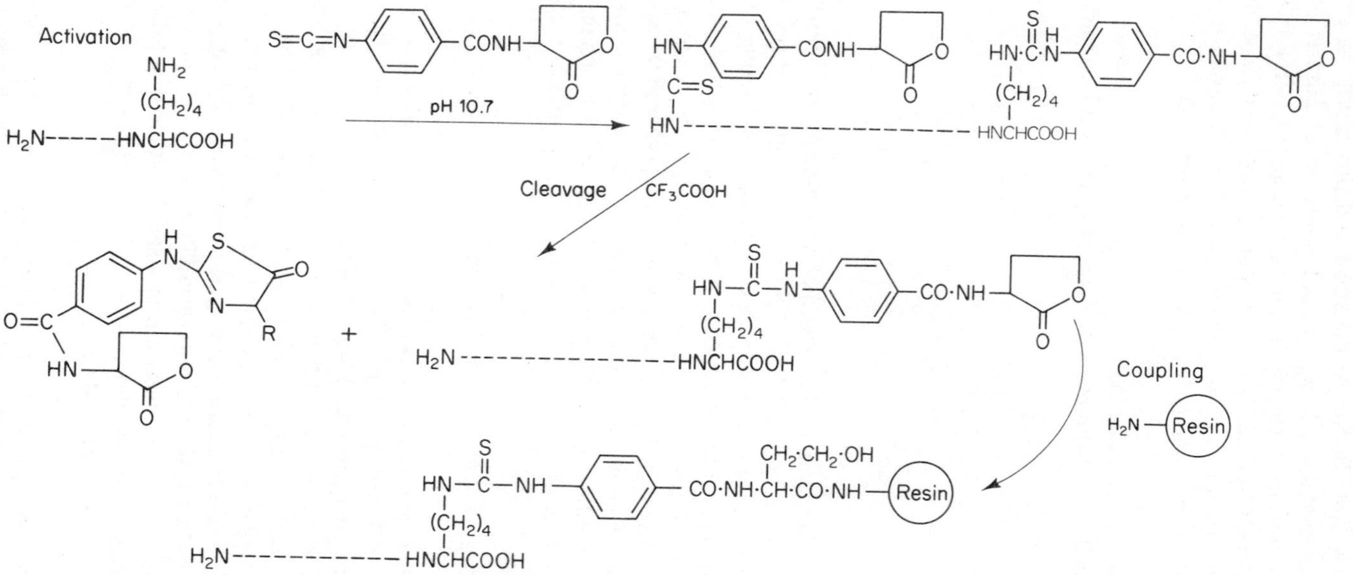

Figure **XXX** Activation and coupling of peptides to an aminated support using N(*p*-isothiocyanatobenzoyl) DL-homoserine lactone

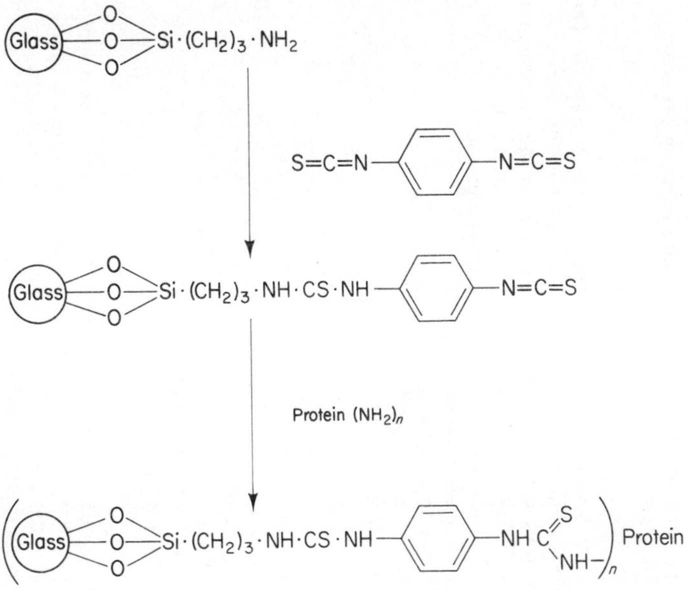

Figure XXXI Coupling of proteins to a porous glass support previously activated with p-phenylene diisothiocyanate

Figure XXXII). The strategy in this case was again to activate the matrix rather than the ligand.

The activated glass support was prepared by suspending amino propyl-glass in a solution of ethylacetate containing iodoacetate and dicyclohexylcarbodiimide for approximately 2 hours. The protein was attached by simply stirring it in 8 M urea solution in the presence of derivatized glass. Finally, when attachment was complete, excess iodoacetamide glass was destroyed by adding mercaptoethanol.

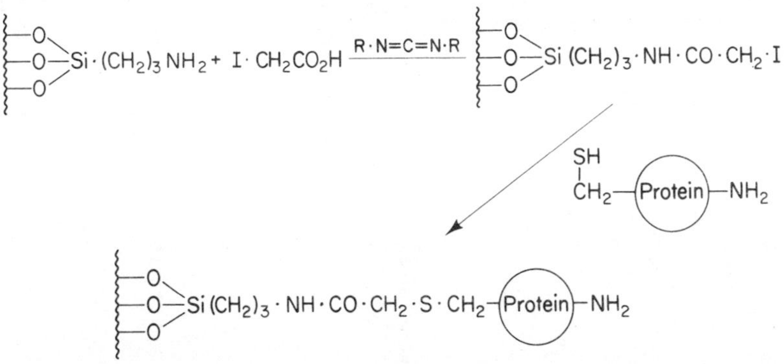

Figure XXXII Attachment of proteins to 3-amino propyl glass previously activated by iodoacetate

Table XIX. Design features of three commercially available solid-phase sequencers

	Sequemat	LKB	Socosi
Reagent storage system	Reagent reservoirs consist of glass vessels with special adaptors that permit easy refilling. The reagents are maintained under nitrogen.	Reagents are contained in five amber coloured bottles and are maintained in an oxygen-free atmosphere.	There are five reservoir vessels under nitrogen pressure.
Pumping system	Milton-Roy mini pumps are used for dichloroethane and methanol. For TFA and PITC Harvard Apparatus infusion-withdrawal pumps are used fitted with 5 ml Hamilton gas-tight syringes.	Five pumps are employed, one for each reagent/solvent. The body of each pump is constructed of glass, with seals of PTFE and a sapphire plunger. The delivery rate is variable. They are mounted together with their motors on a pull-out tray that makes them accessible for easy maintenance.	Pumps are not employed. The reagents/solvents flow throughout the instrument under inducement of nitrogen pressure.
Reagent valves	Pneumatically actuated valves made of either Kel-F or Tefzel, or Delrin. (Delrin is not used near TFA which rapidly attacks it.)	Reagent valves are actuated by N_2 pressure set at 65 lb/sq. in and constructed of Kel-F, or Delrin/Kel-F. A pressure sensor in the N_2 line monitors the actuating pressure and in the event of a pressure drop the programme is stopped until the pressure returns to normal.	
Fraction collector	Constructed of stainless steel with two concentric rows each having a capacity for 24 tubes. This allows thiazolinones from two different peptides to be collected simultaneously. It is enclosed in a stainless steel box thus avoiding the possibility of corrosion by TFA vapours. It is not refrigerated.	This is capable of collecting up to 48 samples, i.e. two rows of 24 tubes permitting collection from each column. It is coated with an acid-resistant polymeric substance to ensure long life. It is maintained in an atmosphere of nitrogen.	The fraction collector is maintained in a nitrogen atmosphere and is refrigerated.

PERSPECTIVES

Initially the solid-phase sequencer may be seen as a cheap alternative to the costly liquid-phase machine, as it is approximately one-fifth of the capital cost. Furthermore, regardless of the initial outlay, when it is compared with the spinning-cup sequencer it is seen that it has the further advantage that the highly refined reagents which are vital for the success of this latter instrument are not neccessary. It is pertinent to point out that not only are the maintenance costs less, but that, theoretically, higher sequencing yields are possible with the solid-phase machine, since entirely organic solvents may be used throughout. The possibility of obtaining stepwise yields greater than 99%, leading to extended sequence runs, exist but has not yet been achieved. Possibly this reflects the reluctance of many investigators to adopt this method of sequencing on account of the involved chemistry of the coupling process. Although considerable research has been devoted to improving the coupling methods, the most expedient route has not yet been conceived and much more exploratory work has yet to be done. From publications of the period 1976–1978 the practical limit for length of sequence deduced by this method is around 25 residues. Employing a dual-column arrangement this could probably be achieved within a single working day, saving many man-hours over the manual alternative. Nevertheless, to justify the large capital cost of this machine, as with all costly instruments, it must be continuously operated, which demands a ready supply of pure peptides. This can only be realized by a group of workers, backed up by an efficient team of technicians, and such organizations are rarely seen in British institutions, at least. This situation probably accounts for the disappointing sales of these machines.

In conclusion, therefore, the solid-phase sequencer is at present limited by the difficult and complicated coupling process, yet it has the potential of being a more useful instrument than the spinning-cup sequencer. Overall it is a cheaper machine to purchase, run and maintain: however, these advantages cannot for the individual worker justify its purchase. This scientist must resist the temptation to elaborate his laboratory and continue to rely on the manual sequencing methods available.

CHAPTER 6

Identification of Amino Acid Phenylthiohydantoins

In general the chemical stability of the phenylthiohydantoin amino acids is very good, which makes them ideal derivatives of the amino acids for identification purposes. During the Edman degradation the thiazolinone is converted to the phenylthiohydantoin (PTH) of the amino acid by acid hydrolysis and is extracted by ethylacetate; however, not all the PTH-amino acids are soluble in the solvent, in particular PTH-Arg, PTH-His and PTH-CySO₃H remain in the aqueous phase.

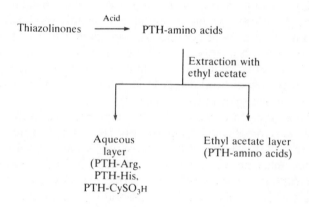

Scheme for extraction of PTH-amino acids.

With the exception of these three, most other PTH-amino acids have poor solubility in aqueous solutions.

All the PTH-amino acids have relatively high melting points and on the whole their thermal stability is good, with the exception of PTH-Ser, PTH-Thr and PTH-Cys. These three phenylthiohydantoins have the common feature of a good leaving group on the β-carbon atom of the side-chain viz. OH, or SH, with the consequence that β-elimination of this group readily occurs forming a dehydrophenylthiohydantoin. Thus dehydration of the PTH of Thr gives rise to 5-ethylidene-

3-phenyl-2-thiohydantoin:

PTH–Thr   PTH–Δ Thr
(λ_{max} 268 nm)   (λ_{max} 325 nm)

This tendency for β-elimination is even stronger with PTH-Ser, the reaction being:

PTH–Ser   PTH–Δ Ser

The PTH dehydroserine formed is very unstable and tends to polymerize giving rise to pink polymers (Ingram 1953). This property has been used to positively identify seryl residues in proteins during Edman degradation with the automatic sequencer, the pink solution being taken as evidence for the presence of a Serine residue (Enfield *et al*. 1975).

As all PTH–amino acids tend to be light-sensitive, particularly PTH–Trp, it is advisable to keep standard PTH-solutions and those derived during stepwise degradation away from direct sunlight. However, after thin layer chromatography, PTH-Trp may be observed as a distinct yellow spot in visible light and this has been used to identify this phenylthiohydantoin after attempts to identify it by GLC had failed (Wunderer and Eulitz, 1978).

(1) THE IDENTIFICATION OF PTH-AMINO ACIDS BY THIN LAYER CHROMATOGRAPHY

This is by far the simplest method for the identification of PTH-amino acids. It is simple, cheap and rapid. It requires no specialist technical skills or equipment, and must be the method of choice for any newcomer to this field. Recent micro-analytical developments have made this method of analysing these amino acid derivatives almost as simple, and quick, as those that have been available for several years for the identification of dansyl-amino acids and it is likely that their impact will have a similar effect in the years to come. Apart from the advantages of speed, cheapness etc., this method has the important advantage of being able to directly identify the amides of glutamic and aspartic acid. This is not possible if the PTH-amino acid is hydrolysed back to the amino acid and subsequently analysed on an amino acid analyser.

In the main the recent advances in this area have been due to the introduction of high-quality manufactured thin layers, of either silica gel or polyamide, on plastic or aluminium foils, and the introduction of double-sided coated thin layers has been particularly beneficial.

Resolution of the PTH-amino Acids by One-dimensional TLC

There is a vast literature describing various solvent systems for the separation of PTH-amino acids and this has recently been reviewed (Rosmus and Deyl 1972). A selection only will be given here, including those methods which the author has found useful in his own research.

The method of Jeppsson and Sjöquist (1967) uses thin layers of silica gel plus a fluorescent indicator (Eastman) and chomatography is performed by the ascending technique in glass tanks lined with Whatman 3MM chromatography paper. The main solvent is system V comprising: heptane/propionic acid/ethylene chloride, (58 : 17 : 25, v/v) and with this system the separation illustrated in Figure XXXIIIa is obtained.

On the whole this is a satisfactory method for PTH-amino acid identification with an over-all sensitivity of about 5 nmol. There are some difficulties associated with the resolution of PTH-Pro, PTH-Phe and PTH-Met; PTH-Asp and PTH-Ser are also poorly resolved; however, this can be improved if the thin layer is rechromatographed in the same solvent. Also, to enhance the separation, particularly of those PTH-amino acids at the lower end of the thin layer, it is re-run in a further solvent, system IV, consisting of: heptane/n-butanol/formic acid, (50 : 30 : 9, v/v), the resulting separation is illustrated in Figure XXXIIIb. A further problem with this system is the separation of PTH-Arg and PTH-His. Jeppsson and Sjöquist suggested heating the thin layer whereupon the PTH of His turns yellow, thus distinguishing it from PTH-Arg. This is unsatisfactory and to overcome this difficulty specific spot tests for PTH-His and PTH-Arg have been applied after the thin layer chromatography. However, as contaminants from the Edman degradation may suppress the colour development of these tests rendering them ineffective and unreliable, this procedure also is not without its snags. To solve this problem Inagami (1973) introduced a method that gave a clear separation of these PTH amino acids, using the

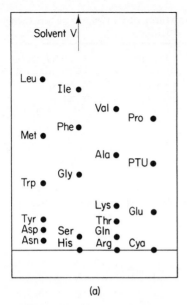

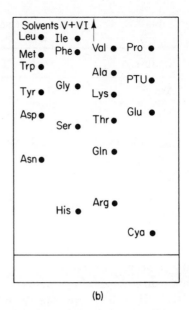

(a)　　　　　　　　　　(b)

Figure XXXIII　(a) TLC separation of PTH-amino acids in solvent system V; (b) TLC separation of PTH-amino acids after chromatography in solvent system V, followed by solvent system IV (Jeppsson and Sjoquist 1967)

solvent system, xylene/95% ethanol/acetic acid, (50 : 50 : 0.5, v/v) (R_f(PTH-His) 0.64; R_f(PTH-Arg) 0.37), and positively identifying the PTH-His after spraying with the Pauly reagent and PTH-Arg with the phenanthraquinone fluorescent stain reagent. The sensitivity by this method for PTH-His is about 2 nmol and that for PTH-Arg 0.05 nmol, compared with the over-all sensitivity of the Jeppsson and Sjöquist method of about 5 nmol.

This procedure for PTH-amino acid identification takes about 2–3 hours before a positive result is obtained, but as the automatic sequencer is able to produce PTH-amino acids awaiting identification at the rate of about one per hour, it is clear that when using the sequencer to its full capacity a bottle-neck situation would soon result at the identification stage. To overcome this quandary Kulbe (1974) introduced a simple, effective and quick method for the simultaneous identification of 10 PTH-amino acids on a single thin layer plate. Kulbe proposed using four standard mixtures (A, B, C, and D) of PTH-amino acids prepared at the concentration of 2 nmol/ml in methanol (Table XX).

Micropolyamide layers, coated on both sides, are used, cut to the dimensions 5 × 5 cm, or 2.5 × 2.5 cm; detection of the PTH-amino acids is made possible by incorporating a fluorescent indicator in the first solvent system, giving a detection limit of between 0.2 and 0.05 nmol. On one side of the layer are placed the standard mixtures A, B and C together with four unknowns and on the reverse side are spotted six further unknowns together

Table XX. Composition of the standard mixtures of PTH-amino acids used in the procedure of Kulbe

Mixture A	Mixture B	Mixture C	Mixture D
Asn	Ala	Asp	Arg
CMC	Ile	Glu	His
Gln	Leu	Gly	CySO$_3$H
Lys	Tyr	Phe	
Met, MetSO$_2$	*N,N-DPhe	Thr	
Pro	*N-Phe	Val	
Ser			
Trp			

*Abbreviations: N,N-DPhe (N,N,diphenylthiourea; N-Phe (N-phenyl-thiourea).

with mixture A. Ascending chromatography is performed in solvent I, comprising toluene/n-pentane/acetic acid, (60 : 30 : 16, v/v) followed by solvent II (25% aqueous acetic acid). The PTH-amino acids poorly resolved in this process, namely PTH-Leu, PTH-Ile, PTH-Thr and PTH-Phe, may be differentiated by doing a further run in solvent III, consisting of 40% aqueous pyridine/acetic acid, (9 : 1, v/v). PTH-His and PTH-Arg may be distinguished by one run in solvent II, the R_f values being 0.89 and 0.83 respectively. This excellent technique makes it possible to identify up to 20 unknown PTHs per hour.

Resolution of the PTH-aminoacids by Two-dimensional TLC

Polyamide thin layer sheets (Cheng-Chin) have been used to identify PTH-amino acids down to a lower limit of 0.05 nmol by incorporating a fluorescent indicator into the thin layer (Summers *et al.* 1973). Double-sided coated thin layers are used, thus allowing the unknown to be run on one side and selected standards on the reverse. This is a rapid method and enables a complete two-dimensional separation of the PTH-amino acids to be obtained within 30 min. The dimensions of the sheet employed are 5×5 cm and the fluorescent indicator is butyl PBD (2-(4'-t-butylphenyl)-5-(4'-biphenyl) 1,3,4-oxdiazole, dissolved in the first solvent. Chromatography is routinely performed in 150 ml glass beakers sealed with a small amount of parafilm. The solvents used are:

solvent I: toluene/n-pentane/acetic acid, (60 : 30 : 35, v/v) containing 250 mg of butyl PBD/litre;
solvent II: 35% aqueous acetic acid.

Separation of 16 of the PTH-amino acids was possible by this method, as illustrated in Figure XXXIV, and this was achieved within 30 min.

This procedure has been slightly modified by Kulbe (1974), who used the

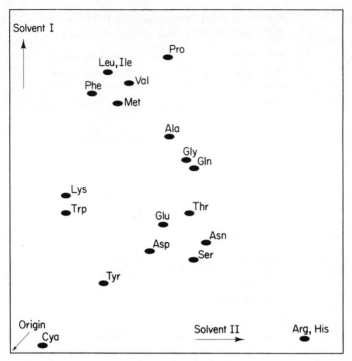

Figure XXXIV Separation of PTH-amino acids by two-dimensional TLC on polyamide layers (Summers *et al.* 1973)

solvents:

solvent I: toluene/n-pentane/acetic acid, (60 : 30 : 16, v/v) containing 250 mg/litre of fluorescent indicator;
solvent II: 25% aqueous acetic acid.
The moderately improved separation gives better resolutions of PTH-Glu and PTH-Gln, PTH-Asp and PTH-Asn; however the pairs PTH-His/PTH-Arg, PTH-Leu/PTH-Ile are still not completely separated and the use of a third solvent is necessary before a positive identification can be made.

Detection of PTH-amino Acids on Thin Layers

The simplest method of detecting PTH-amino acids involves the use of a fluorescent indicator incorporated into the thin layer. The plates may either be purchased with the fluorescent compound already in the layer, or it can, as in the method of Summers *et al.* (1973), be dissolved in the chromatography solvent. After development the PTH-amino acids may be identified by viewing the dry thin layer under a UV light, when they appear as dark areas on a bright fluorescent background.

Table XXI. Colours produced by PTH-amino acids after staining with ninhydrin–collidine reagent

PTH-amino acid	Colour produced
Glycine	Deep orange
Alanine	Red-violet
Serine	Red-violet
Threonine	Pale brown
Valine	No colour
Leucine	Faint grey
Isoleucine	Faint grey
Proline	Pale pink
Glutamic acid	Burnt umber
Aspartic acid	Pink
Glutamine	Brown
Asparagine	Lemon
Histidine	Pale yellow
Lysine	Faint pink
Arginine	Pale yellow
Phenylalanine	Faint yellow
Tyrosine	Lemon
Tryptophan	Yellow

However if there is any doubt as regards the identification, confirmation may be obtained after staining with the ninhydrin-collidine reagent Roseau and Pantel (1969), when the PTH-amino acids have characteristic colours (Table XXI); this innovation was very helpful in studies on the sequence of horse pancreatic ribonuclease (Scheffer and Beintema 1974).

Advantages of TLC

TLC identification of PTH-amino acids, particularly the innovations of Summers et al. (1973) and Kulbe (1974), is of major importance in amino acid sequencing. This has been made possible by the commercial availability of uniform and consistent thin layer sheets. Undoubtedly the principal attraction of this method is its low cost, particularly if the operation is scaled down to use the 2.5 × 2.5 cm layers, when amino acid residue identification is obtained for a fraction of a farthing. Its speed is another bonus and as the sequencer can produce about 20 PTH-amino acids per day, which of necessity must be identified as soon as possible, to avoid any decomposition caused by prolonged standing, it is the only means of keeping up with a machine working at full capacity. The method is also highly sensitive, having about 10 × the sensitivity of GLC, and it is very reliable, but above all its main advantage is in its simplicity—requiring no skilled technical assistance, or elaborate expensive equipment—it is clearly the method of choice.

(2) THE IDENTIFICATION OF PTH-AMINO ACIDS BY GAS LIQUID CHROMATOGRAPHY

The early attempts at separating the PTH-amino acids by GLC used essentially the same approach as that for other natural products (Pisano *et al*. 1962). The column packings used consisted of thin film coatings of very thermostable liquid phases, in conjunction with an argon ionization detection system. The initial results were satisfactory: however, difficulties were experienced with the phenylthiohydantoin derivatives of Ser, Thr, Asn, Gln, His and Arg. The phenylthiohydantoins of both Ser and Thr had undergone dehydration, and the derivatives of Gln and Asn had also been modified. It was also found that the PTH derivatives of the basic amino acids were not sufficiently volatile for analysis by this method. These initial stumbling-blocks were overcome by the introduction of new thermally stable and more polar polysiloxane liquid phases, together with the use of more powerful silylating reagents which were found to convert the relatively non-volatile and unstable PTH-amino acids to trimethylsilyl derivatives having satisfactory chromatographic properties. As yet, however, no satisfactory solution to the inherent involatility of PTH-Arg has been found, and this amino acid derivative cannot be identified by GLC analysis. Because of this, GLC analysis of PTH-amino acids cannot be considered as a complete method, in itself, and must always be used in conjunction with other analytical systems.

The procedure employed in most sequence laboratories is that originally described by Pisano *et al*. (1972). These workers considered the phenylthiohydantoin amino acids to fall into three categories, differing markedly in their volatilities and compatibilities with different liquid phases: namely, Groups I, II and III (Table XXII).

Group I includes the PTHs that are most volatile and in general give symmetrical peaks. Group II derivatives are the least volatile and are eluted long after those of Group I (with the exception of PTH-Trp) and tend to give unsymmetrical peaks due to interactions with the column packings. The

Table XXII. PTH-amino acids differing in gas chromatographic behaviour

Group I	Group II	Group III
PTH-Ala	PTH-Asn	PTH-Asp
PTH-Gly	PTH-Gln	PTH-CMC
PTH-Val	PTH-Tyr	PTH-CySO$_3$H
PTH-Leu	PTH-His	PTH-Glu
PTH-Ile	PTH-Trp	PTH-Lys
PTH-Met		PTH-Ser
PTH-Pro		PTH-Thr
PTH-Phe		

PTH-derivatives included in Group III must be modified by silylation before GLC analysis is attempted.

It is therefore clear that because of this spectrum in the properties of the PTH-amino acids it is not feasible to analyse for all possible ones in a single GLC analysis run, and in general it is necessary to analyse each residue derivative from the steps of the Edman degradation at least twice. Considering that one GLC run takes approximately 40 min, two runs would take about 2 hours taking into account any down-time. When working with the automatic sequencer about 20 residues per day are prepared for analysis and it is thus clear that a bottleneck of samples awaiting analysis would soon materialize if GLC was the only method of analysis employed. There are a few affluent laboratories that deploy two GLC machines to cope with this problem, one for analysis of the unmodified PTH-derivative, and the other for the silylated PTH-amino acid. An alternative less expensive approach has been to silylate all samples regardless of whether the amino acid derivative might be a member of Group I. However, as pointed out by Pisano et al. (1972) the identification of the majority of PTH-amino acids is not facilitated by silylation, for they found that although individually all PTHs would react quantitatively with the silylating agent, the rate of reaction and stability of the derivatives was highly variable, this being particularly true for phenythiohydantoins of Asn and Gln. Occasionally Pisano and co-workers observed no peaks after silylation, whereas before silylation reasonable responses had been observed. Silylation also has the disadvantage that many additional peaks are observed in the chromatogram that were not present in the unsilylated product; though fortunately none of these artefacts appear to coincide with the known retention times of any of the PTH-amino acids. Wunderer and Eulitz (1978) found that PTH-Ser on silylation gave two or three peaks and that this made quantitation impossible.

Tomita et al. (1978), in their studies on glycophorin A, one of the principal intrinsic transmembrane proteins of the human erythrocyte, adopted the compromise of identifying the PTH derivatives of Ala, Thr, Ser, Gly, Val, Pro, Leu and Ile by direct GLC analysis and a further eight PTH-amino acids (Met, Phe, Asp, Glu, Asn, Gln, Lys, Tyr) after silylation. The four remaining PTH amino acids, viz. PTH-Cys, PTH-Trp, PTH-His and PTH-Arg, were identified using other analytical procedures. Even with this approach difficulty was experienced in resolving PTH-Leu/Ile and in obtaining a satisfactory response from PTH-Asn and PTH-Gln. However, Bennett et al. (1978), in their studies on dihydrofolate reductase, found it possible to resolve PTH-Ile/Leu by using a column of 1.5% AN-600 instead of the usual 10% DC-560, and Closset et al. (1978) found they could obtain satisfactory resolution of this pair of phenylthiohydantoins by silylating the sample before analysis.

Reagents for Silylation

The usual reagent for the silylation of PTH-amino acids is either N,O-bis(trimethylsilyl) acetamide (BSA), or N,O-bis(trimethylsilyl) trifluoroacetamide (BSTFA).

$$
\begin{array}{cc}
\underset{|}{Si(CH_3)_3} & \underset{|}{Si(CH_3)_3} \\
\underset{|}{O} & \underset{|}{O} \\
CH_3-C=N-Si(CH_3)_3 & CF_3-C=N-Si(CH_3)_3 \\
\textit{BSA} & \textit{BSTFA}
\end{array}
$$

Both of these reagents are powerful trimethylsilyl donors and react to replace labile hydrogen in the amino acid derivatives with a $Si(CH_3)_3$ group. Thus the PTH-amino acids become N- and O-trimethylsilyl derivatives. The particular advantage of BSTFA over BSA is the greater volatility of its by-products monotrimethylsilyltrifluoroacetamide and trifluoroacetamide than that of the unfluorinated by-products of BSA. The by-products of BSTFA will usually elute well before any of the silylated amino acid derivatives.

A further silylating agent is N-methyl-N-trimethyl-silylhepta-fluorobutyramide, which has been used by Beyreuther *et al.* (1975) to

$$
CF_3CF_2CF_2-C \overset{\displaystyle O}{\underset{\displaystyle \underset{|}{N}-Si(CH_3)_3}{\diagup}}
$$
$$
\overset{|}{CH_3}
$$

derivatize the PTHs of Glu, Asp, Ile and Tyr in their studies of the sequence of *lac* Repressor from *E. coli*.

Method of Silylation

Silylation of the PTH-amino acid may be carried out by mixing it in ethyl acetate with an equal volume of BSA in either a conical glass test tube, or equivalent plastic tube, of 0.5 ml capacity, sealed with a PTFE-lined rubber septum. The reaction vessel is heated in a water bath for 15 min at 50°C and the samples withdrawn directly through the septum with a 10 μl Hamilton syringe (Pisano *et al.* 1972).

However, the most successful method of silylation is the 'on-column' derivatization. This is usually performed by injecting the reagent with the sample into the column at an inlet temperature of about 250°C, the best results being obtained by filling the syringe first with the silylating agent and then the PTH-amino acid. The reverse sequence of operations has been found to result in incomplete silylation, which shows up as double peaks for most of the PTH derivatives (Beyreuther *et al.* 1975).

General Procedure

The usual support is Chromosorb W (100–120 mesh) with a 10% stationary phase of either DC 560, or its equivalent SP 400; the former having been successfully used by amongst others, Tomita *et al.* (1978), and Butler *et al.* (1977), and the latter by Beyreuther *et al.* (1975), Bennett *et al.* (1978), and Butkowski *et al.* (1977). As metal columns cannot be used—they cause destruction of the PTH-amino acids—the usual column material is glass with dimensions of about 2 mm × 3 ft, and before use it is treated, together with the glass wool plug and the support, with dichlorodimethylsilane in toluene. This reduces the number of adsorptive sites, which tend to cause tailing of the samples. Before analysis is begun the column is conditioned at 300°C for 16 hours: however, analysis is performed with a column temperature somewhere between 165° and 300°C, the optimum temperature being determined by the investigator. Normally the carrier gas is nitrogen, at a flow rate of about 65 cm^3/min. If helium is used instead, better resolution is obtained (Figure XXXV), at a small increase in cost.

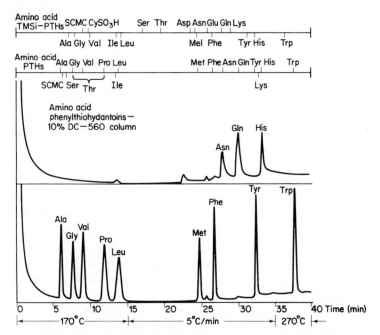

Figure XXXV   Separation of amino acid PTHs on a 4 ft × 2 mm, 10% DC-560 column. Sample sizes: 1 μg each PTH-Ala, Gly, Val, Pro, Leu, Met, Phe, and Tyr; 2 μg PTH-Trp, Asn, Gln, and His. Helium flow 65 ml/min. (From Pisano *et al.* (1972), reproduced by permission of Academic Press Inc.)

Advantages of GLC Analysis

The principal advantage of GLC analysis of PTH-amino acids is that it is quantitative, and repetitive yields may be calculated for most stages of the Edman stepwise degradation. Moreover interest is centred on this approach as it offers a possible means for the automated analysis of PTH-amino acids resulting from automated protein degradation. Automatic sample injectors are already on the commercial market and have given encouraging results with standards (Pisano *et al.* 1972). Samples may be analysed every 50 min and at this rate it seems possible that it could keep pace with the sequencer, even after including a silylation step. Nevertheless, there still remain several problems of derivatization to be solved before this costly aim is a reality.

Difficulties and Disadvantages of GLC Analysis

The major problems associated with this method of analysis are not related to the technology but to the chemistry of successful derivatisation. Although there are many reports in the literature of complete separations of most of the PTH-amino acids, when standards are used as samples, it is clear from the literature from sequence laboratories that GLC analysis, in the main, is used to analyse for at the most 16 of the 20 possible PTH derivatives. The four difficult samples are PTH-His, PTH-Arg, PTH-Cys and PTH-Trp. No successful method for the derivatisation of PTH-Arg has yet been devised and this surely deserves more attention than it has yet received. Some reports have indicated successful analysis of PTH-Cys, yet others report difficulties (Hogg and Hermodson 1977) and the same applies to PTH-Trp (Wunderer and Eulitz 1978). Even a stable derivative of cysteine, namely PTH-CMC, undergoes degradation during GLC analysis, giving rise to amongst other products PTH-dehydroserine which rapidly polymerises; however, PTH-pyridylethylated cysteine has been successfully analysed by GLC using an isothermal program at 220°C (Mak and Jones 1976).

Apart from these problems, which are of no mean significance, the other drawback to adopting this method for routine PTH-amino acid analysis is that it is beyond doubt a highly skilled technique requiring not only competent technical assistance but also expensive capital equipment. Also, it is not as quick as other available means of analysis and in a 8 hour working day it is not possible to identify all the PTH's derived from an automatic sequencer during the previous 24 hours. At present therefore, GLC analysis does not offer the many advantages that other methods of analysis have.

(3) IDENTIFICATION OF PTH-AMINO ACIDS BY HIGH-PRESSURE LIQUID CHROMATOGRAPHY (HPLC)

The relatively new technique of HPLC is now frequently being used to identify PTH-amino acids. It is a very sensitive and rapid method of analysis, but it does require expensive equipment and technical assistance. Nevertheless, it has been applied in recent years to several proteins, including: human prethrombin 2 (Butkowski *et al.* 1977); brain S-100 protein (Isobe *et al.* 1977a); bacteriophage T_4 internal protein I (Isobe *et al.* 1977b); dihydrofolate reductase from mouse sarcoma and an *E. coli* mutant (Rodkey and Bennett 1976, Bennett *et al.* 1978); and chick skin collagen (Dixit *et al.* 1977).

Although the resolution obtained is not as good as that obtained with GLC, HPLC has the following advantages:

(1) it is not necessary to prepare volatile derivatives;
(2) short analysis time—in some cases as little as 6 min;
(3) it is very sensitive and can go down to 50 pmol if a suitable gradient system is used;
(4) using this method it is possible to differentiate between PTH-Leu/Ile;
(5) it has the potential for automation.

Several different instruments have been employed for the analysis of PTH-amino acids, e.g. the Du Pont Model 830 liquid chromatograph, the Waters Associates ALC/GPC 202 high-pressure chromatograph etc. Each of these instruments has the essential features of a column, a reliable pump, a gradient maker capable of giving reproducible results, and a UV detection monitor coupled to a recording device.

Various stainless steel columns have been employed with diameters between 2 mm and 4 mm and lengths 15 cm to 500 cm, and may be purchased pre-packed with support, or may be packed dry with stationary phase in the laboratory. Supports that have been used include Partisil 5, or 10 (Reeve Angel), and μ Bondapak C_{18} (Waters Associates), with eluting solvents of diethylmethane containing 0.32–0.33% methanol, and acetonitrile–sodium acetate (pH 4.0) respectively. The PTH-amino acids are detected by their UV absorbance using a flow-through cell and UV monitor connected to a recording device. The PTH's are characterized by their retention times.

In general the PTH-amino acids fall into three separate groups depending upon their individual polarity:

Non-polar	*Relatively non-polar*	*Highly polar*
PTH-Pro	PTH-Trp	PTH-Cya
PTH-Leu	PTH-Gly	PTH-Glu
PTH-Ile	PTH-Lys	PTH-Asp
PTH-Val	PTH-Tyr	PTH-His
PTH-Phe	PTH-Thr	PTH-Arg
PTH-Met	PTH-Ser	
PTH-Ala	PTH-Asn	
	PTH-Gln	

Isobe and Okuyama (1978) were able to separate and identify all the PTH-amino acids except PTH-Arg, PTH-His and PTH-Cya by using two columns, one for the apolar and the other for the polar PTH's, in their studies on the S-100 protein from brain tissue.

The analysis of PTH-amino acids by HPLC is not without its difficulties, namely:

(1) obtaining reproducible retention times;
(2) the interference caused by UV-absorbing impurities resulting from the Edman degradation.

However, regardless of these minor difficulties this technique has an assured position in the armoury of methods for PTH-amino acid analysis. Firstly because it is the most likely candidate to be fully automated and linked to a sequencer, and secondly it is extremely useful in 'microsequencing'. In this technique the radioactive PTH-amino acid from the sequencer is mixed with non-radioactive PTH-amino acid standards and the mixture fractionated by HPLC. The effluent after passing through the UV detecting system is mixed with scintillation fluid and passed through a scintillation flow-cell. The output from the counter can then be displayed directly on the HPLC trace giving immediate identification of the radioactive amino acid.

(4) THE IDENTIFICATION OF PTH-AMINO ACIDS BY THEIR CONVERSION BACK TO AMINO ACIDS

This is probably the simplest procedure for the identification of PTH-amino acids and may also be applied to the precursor amino acid anilino-thiazolinones with a consequent saving in time. As both the equipment and expertise are readily available in all laboratories involved in amino acid sequencing it is clear that this approach has considerable advantages over other methods, and with the advent of rapid and highly sensitive analysers the capacity to maintain pace with the automatic sequencer. It is also conceivable that the thiazolinones from the sequencer could with the appropriate technical innovations be fed through a hydolysis coil to an automatic amino acid analyser, so perfecting the automatic sequencing of proteins. It is unfortunate that so little work has been done to improve the hydrolytic conditions that at present prevent this concept becoming a feasible and successful proposition.

The first serious study of the conditions necessary for the hydrolysis of PTH-amino acids was by van Orden and Carpenter (1964). They studied the hydrolysis of PTH's using either 0.1 N NaOH or 6 N HCl, and gave details of the conditions necessary for both of these reagents to back-hydrolyse most of the PTH-amino acids in reasonable yields. Several years elapsed before Smithies et al. (1971) made a detailed study of this problem and introduced two new reagents, viz. 57% HI and NaOH–dithionite. These investigators applied the new reagents to quantitate the steps in the automatic Edmon

Table XXIII. Methods for the hydrolysis of PTH-amino acids

Method	Details	Reference
Acid hydrolysis	(1) 6 N HCl, 150°C, 24 hours	van Orden and Carpenter (1964)
	(2) 57% HI, 127°C, 20 hours	Smithies et al. (1971)
	(3) 4 M CH$_3$SO$_3$H, 150°C, 4 hours	Mendez and Lai (1975)
	(4) 5.7 M HCl (+0.1% w/w SnCl$_2$) 150°C, 4 hours	Mendez and Lai (1975)
	(5) 55% HI, 130°C, 18 hours	Tomita et al. (1978)
Alkaline hydrolysis	(1) 0.1 N NaOH, 120°C, 12 hours	van Orden and Carpenter (1964)
	(2) 0.1 M Na$_2$S$_2$O$_4$ in 0.2 M NaOH, 127°C, 3.5 hours	Smithies et al. (1971)

Table XXIV. Percentage recoveries of amino acids after hydrolysis of their PTH-derivatives

Amino acid	57% HI (1)[i]	47% HI (3)	6 N HCl (2)	5.7 M HCl (3)[h]	4 M CH$_3$SO$_3$H (3)	0.2 N NaOH (1)[j]	0.1 N NaOH (2)	0.01 N NaOH (3)[k]
Asp	80	100	88	97	103	80	89	78
Thr	90[b]	89[b]	—	90[b]	18[b]	1	67[c]	20
Ser	50	1	—	68[d]	9	5[d]	—	1
Glu	80	100	81	97	100	50	66	23
Pro	70	78	88	94	91	70	96	94
Gly	80	95	104	99	99	80	96	118
Ala	90	100	78	99	100	80	86	102
Cys	60[e]	ND	ND	ND	ND	5[e]	ND	ND
Val	70	72	88	61	69	13	98	96
Met	—	—	67	92	75	90	84	76
Ile[f]	70	65	90	60	68	100	100	87
Leu	90	82	59	82	81	90	71	83
Tyr	60	80	76	67	70	100	97	92
Phe	80	72	86	70	72	100	95	88
His	20	100	ND	99	93	20	70	48
Trp	60[a]	ND	ND	97[a]	ND	40	ND	ND
Lys	70	100	ND	92	84	30	72	30
Arg	30	81	ND	99	85	5[g]	53[g]	44

References: (1) Smithies, O., Gibson, D., Fanning, E. M., Goodfliesh, R. M., Gilman, J. G., and Ballantyne, D. L. (1971) *Biochemistry* **10** 4912–4921. (2) van Orden, H. O. and Carpenter, F. H. (1964) *Biochem. Biophys. Res. Commun.* **14** 399–403. (3) Mendez, E. and Lai, C. Y. (1975) *Anal. Biochem.* **68** 47–53.
Footnotes: ND, not determined; a, Trp was determined as Gly + Ala; b, Thr was determined as α-aminobutyric acid; c, Thr was determined as Gly; d, Ser was determined as Ala; e, Cys was determined as Ala; f, Ile includes the value for alloisoleucine; g, Arg was determined as ornithine; h, Contains 0.1% w/w of SnCl$_2$; i, the minimum recovery only is quoted; j, containing Na$_2$S$_2$O$_4$; k, containing 0.1% β-mercaptoethanol.

degradation. More recently, Mendez and Lai (1975) have introduced several new hydrolytic reagents and conditions, in particular describing the finding that 6 N HCl in the presence of $SnCl_2$ gives much improved recoveries than the acid alone. To date all these methods have been used in various sequence studies and the individual conditions are given in Table XXIII. The recoveries obtained from the different methods are summarized in Table XXIV.

All of the available methods have been found to give good yields of the following amino acids: Asp, Glu, Pro, Gly, Ala, Val, Leu, Tyr and Phe. The PTH's of Asn/Gln are always converted to the respective acids and this is one unsatisfactory aspect of this method; the location of the amides must then depend upon another method of identification. Isoleucine is always recovered as a mixture of the diastereoisomers Ile/allo-Ile; the latter isomer, having different chemical properties, is found to elute on ion-exchange chromatographic analysis in a different position than Ile, in fact allo-Ile is eluted slightly before Ile. Although no detailed study has been made on this isomerization it is thought to involve the dehydro intermediate:

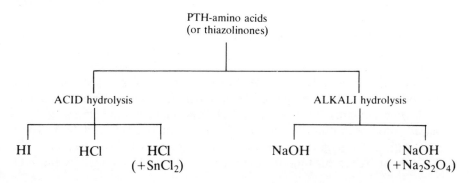

The derivatives of Trp and Cys are always destroyed, nevertheless attempts have been made to identify them by the artefacts formed on back-hydrolysis, namely Gly and Ala, from Trp, and Ala from Cys. The PTH of serine also in general gives poor recoveries and the corresponding derivative of Thr also gives rise to the artefact α-aminobutyric acid, or glycine, which may be used for its identification and quantitation.

Methods for the Back-hydrolysis of PTH-amino Acids

(1) 6N HCl

In many instances this method is used to complement another method of identification, for example if TLC was the main method of identification, back-hydrolysis using 6 N HCl has been used to differentiate between PTH–Leu and PTH-Ile (Begg *et al.* 1978). The amino acids destroyed by this method are Thr, Ser, Cys, and Trp. Methionine is recovered in satisfactory yield. Mendez and Lai (1975) found that the yields may be increased by the incorporation of $SnCl_2$ in the hydrolysis solution, also this innovation enabled Thr to be quantitated as α-amino-butyric acid and serine as alanine. This approach has been used by Beyreuther *et al.* (1978) in their study of citrate-lyase acyl carrier protein.

(2) 57% HI

This has probably been the most widely used method, various concentrations of HI being employed in the range 44–57%, without any significant differences. The following modifications are observed with this procedure.

$$PTH - Asn/Gln \longrightarrow Asp/Glu$$

$$\left. \begin{array}{l} PTH-CMC \\ \\ PTH-Ser \end{array} \right\} \longrightarrow Ala$$

$$PTH - Trp \longrightarrow Gly \ and \ Ala$$

$$PTH - Thr \longrightarrow \alpha\text{-aminobutyric acid.}$$

If these artefacts are determined, the only amino acids not identified by this method will be Met and Cys, and in their study of the structure of Glycophorin A, Tomita *et al.* (1978) took the absence of any identifiable amino acid, at any step of the sequence degradation, after hydrolysis of the PTH-amino acid with HI, to be indicative of glycosylation.

Although PTH-Pro is hydrolysed in good yield to proline, some side reactions do occur giving rise to small amounts of artefactual material eluting on amino acid analysis near to histidine.

In general the approach is to use either the thiazolinone solution, or to combine both the ethylacetate and aqueous phases if conversion has been performed, and hydrolyse in sealed evacuated hydrolysis tubes or in open test tubes (50×10 mm) placed in a small desiccator (10 cm diameter). Each tube contains 0.1 ml of 57% HI, and an additional 25 ml of HI is placed into the bottom of the desiccator which is evacuated (0.01 Torr), fixed with a clamping ring and heated in a normal hydrolysis oven at 130°C for 20 hour. After this time the tubes are dried over NaOH pellets and P_2O_5

under high vacuum at 65°C. The residues are then ready for amino acid analysis. To quantitise each sample 50 nmol of PTH-norleucine is added prior to hydrolysis.

One of the problems associated with this method is that Ala may result from the degradation of at least three different precursors, namely PTH-Ala, PTH-CMC and PTH-Ser. To ensure positive identification the approach must be supplemented by another method. Closset *et al.* (1978), in their study of the sequence of the β-subunit of porcine follitropin, distinguished these three derivatives by the following criteria: the alanine derivative was determined by gas chromatography and by a high yield of free alanine after back hydrolysis and amino acid analysis. The CMC derivative was identified by its intrinsic radioactivity introduced by the use of ^{3}H-labelled iodoacetate for the alkylation of the cysteine residues in the protein. Finally CMC and the seryl derivative were also characterized by a low yield of free alanine on the amino acid analyser after back-hydrolysis.

(3) NaOH–dithionite

This method enables Met and Trp to be recovered unaltered, Met in yields close to the theoretical and Trp in about 50% yield. Alanine is recovered, but Ser and Cysteine (or CMC) are destroyed. Threonine is almost completely destroyed but gives a little α-aminobutyric acid. Arginine is also destroyed but a small amount of ornithine is recovered. Hydrolysis is performed in tubes placed in a desiccator with extra NaOH–dithionite solution placed at the bottom—this prevents the samples drying out. To each sample tube is added 0.2 ml of 0.1 M solution of $Na_2S_2O_4$ (sodium dithionite) in 0.2 M NaOH. The desiccator is sealed and heated in an autoclave at 127° at 21 lb/sq. in pressure. Following hydrolysis the samples are acidified by adding 20 μl of 3 M HCl containing 10% thiodiglycol (TDG), however it has been found that following this procedure (Glu/Gln) and (Asp/Asn) give other ninhydrin–reacting species that appear on the analyser near to the positions of Val and Pro respectively. If the samples are acidified in the presence of dithiothreitol (10%) instead of TDG more satisfactory results are obtained (Jabusch *et al.* 1978).

The combined use of both acid and alkaline hydrolysis is a satisfactory approach and was used by Stone *et al.* (1977) in their studies of the sequence of dihydrofolate reductase. Back-hydrolysis was routinely performed using 57% HI followed by hydrolysis with 0.1 M $Na_2S_2O_4$ in 0.2 M NaOH in those cases where the hydriodic procedure failed to give an unequivocal identification.

CHAPTER 7

Manual Methods of Amino Acid Sequencing

DETERMINATION OF N-TERMINAL RESIDUES AND SEQUENCES

The determination of amino terminal residues in both peptides and proteins is usually one of the first tasks performed in any sequence study. The finding of a unique amino terminal residue is good confirmation that the peptide/protein is as pure as hoped for. Two chemical approaches, that were devised during the period 1945–1950, are the main means of tackling this problem. The first involves labelling the amino terminal residue covalently with a suitable group that is stable to subsequent acid hydrolysis. The group is chosen for its ability to be detected by either colour, fluorescence, or radioactivity, and identification is made after comparing the product, after acid hydrolysis, with standard amino acid derivatives. This approach was first used by Sanger (1945), who employed 2,4-dinitrofluorobenzene for this purpose. This was the technique he later used in his classic work on the structure of insulin. The use of dinitrofluorobenzene was a brilliant piece of intuitive chemical innovation that was to initiate the subsequent rapid development in protein structure determination. However, at the present time this reagent is not used as it has been replaced by superior reagents (see Figure XXXVI).

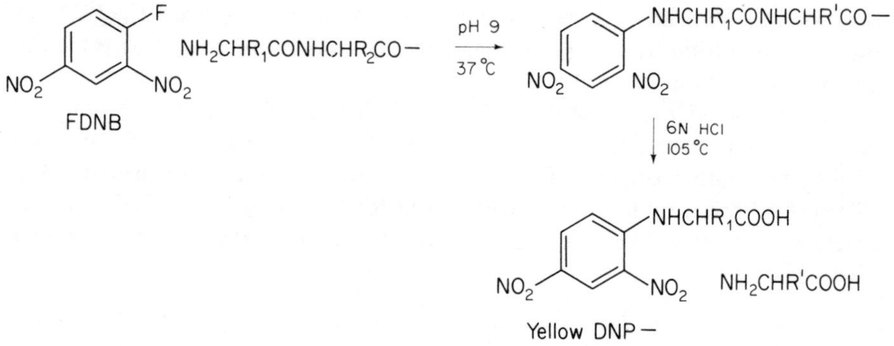

Figure XXXVI Use of Sanger's reagent to label the N-terminal residue of a peptide

102

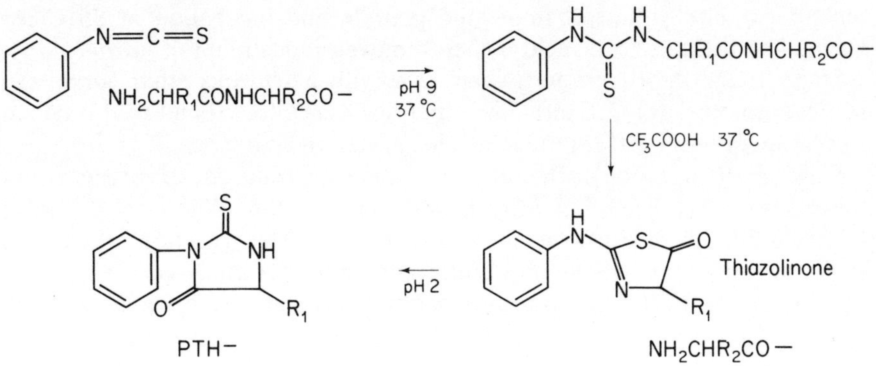

Figure XXXVII   The Edman degradation reaction

The second approach was that devised by Edman in 1950, the procedure now bearing his name. The reagent used is phenylisothiocyanate, and this reacts at alkaline pH with the amino group of the N-terminal residue. The resulting N-phenylthiocarbamoyl derivative is cyclized in the presence of acid to give the thiazolinone corresponding to the N-terminal residue, and a peptide having one residue less than the original (Figure XXXVII). The thiazolinone is unstable and cannot be identified as such, however it is readily converted to the more stable phenylthiohydantoin in the presence of acid.

These two procedures are the mainspring from which all of the presently available manual sequencing methods have been derived. These methods will now be considered in detail.

(1) The Edman Method of Sequence Determination

(i)   The Edman Degradation

There have been many modifications of the simple reaction sequence originally proposed by Edman. One of the most successful variations has been that devised by Peterson *et al*. (1972) and this is outlined below.

Stage 1 (coupling: the peptide (100 nmol) is dissolved in 0.2 ml of 0.4 M dimethyl-allylamine-trifluoroacetic acid buffer, pH 9.5, in 1-propanol–water (3 : 2 v/v). Phenylisothiocyanate (10 μl) is added under N_2 and the solution mixed. It is then incubated at 55°C for 20 min.

Stage 2 (washing): to the reaction mixture is added 1 ml of benzene, the solution mixed and centrifuged. The upper benzene layer is discarded, and the aqueous solution, after drying, is extracted with ethyl acetate (0.5 ml) under nitrogen. The ethyl acetate layer is discarded and the aqueous solution reduced to dryness.

Stage 3: cleavage is performed on the addition of 100 μl of trifluoroacetic acid to the phenylthiocarbamoylated peptide and incubation at 55°C for 8 min. The trifluoroacetic acid is then removed in a stream of nitrogen.

Stage 4: the dried residue is extracted with anhydrous ether containing 10^{-4} M dithiothreitol (1.0 ml). The ether layer is collected and to it is added a small quantity of PTH-norleucine as an internal standard.

Stage 5 (conversion): the ether solution is reduced to dryness and conversion to the PTH-derivative is performed by the addition of 0.2 ml of 1 N HCl and incubation at 80°C for 10 min. This reaction mixture is extracted with ethyl acetate and both layers taken for analysis.

A single cycle of this procedure takes approximately 50 min, and the

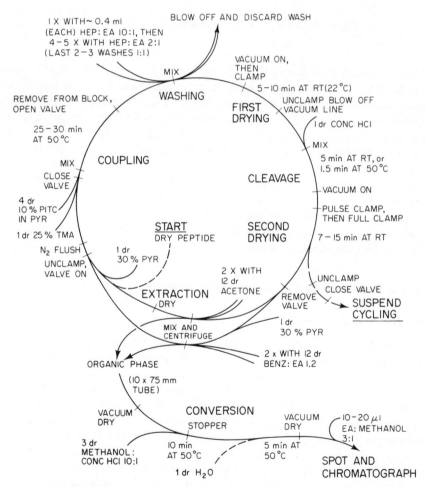

Figure XXXVIII The Edman cycle as proposed by Tarr (1975). Abbreviations: dr, drop; TMA, trimethylamine; PITC, phenylisothiocyanate; HEP, heptane; EA, ethyl acetate; RT, room temperature; BENZ, benzene. (Reproduced by permission of Academic Press Inc.)

amino acid sequence of up to 20 residues may be routinely determined using less than 100 nmol of peptide. van Eerd and Takahashi (1976) introduced an important modification of this reaction: they added water directly to the residue after removal of the trifluoroacetic acid whereupon they extracted the thiazolinone with ethyl acetate. This was found to reduce mechanical losses of the peptide at each step of the Edman.

Walker *et al.* (1977) introduced the technique of 'double coupling', in which a further 10 µl of phenylisothiocyanate was added after 30 min of the coupling step. This procedure was used with peptides having a high proline content, as it is known that proline residues couple with some difficulty in the Edman procedure.

With hydrophobic peptides solubility problems can be overcome by the use of Braunitzer reagents, as is carried out in automatic liquid-phase sequencing. van Eerd and Takahashi (1976) used Braunitzer reagent 1 (4-sulphophenyl isothiocyanate) in place of phenylisothiocyanate in studies of tryptic peptides from ox cardiac troponin C. These authors report excellent accomplishments with the manual Edman degradation that enabled them to sequence the whole of this protein using only 60 mg. This efficient degradation enabled them to sequence up to 36 residues of one tryptic peptide. These results are certainly encouraging for those laboratories that lack the presence of an automatic sequencer.

Tarr (1975) has also introduced important modifications of the Edman degradation that enables up to 40 residues of a peptide to be sequenced manually using between 5 and 50 nmol of peptide. Tarr considered that the primary defect of the classical Edman degradation was due to desulphurization of the peptide by atmospheric oxygen and to this end he included ethanethiol in the coupling buffers to ensure reducing conditions. His reaction cycle is illustrated in Figure XXXVIII. Other modifications were the use of a micro-reaction vessel that enable the introduction and removal of reagents under a continuous nitrogen barrier. Also after coupling the washing procedure was improved by the use of heptane:ethyl acetate mixtures instead of benzene, and in particular cleavage was performed using concentrated HCl since Tarr considered that it is the trifluoroacetic acid that leads to the formation of accumulated 'oil' in the conventional Edman reaction. He was able to sequence manually 39 residues of a 44-residue peptide using this improved procedure, and pointedly commented that the actual length of sequence determination achieved was solely limited by the 'patience of the investigator'. Other successes with the manual Edman sequence degradation are summarized in Table XXV.

(ii) The 'Subtractive' Edman Method

The 'subtractive' Edman is only applicable to small peptides, as it depends on the difference in analysis before and after each step of the degradation.

Table XXV. Some achievements at determining long sequences using manual methods

Sample	Amount used (nmol)	No. of residues sequenced	Method used	Reference
Peptide	<50	39	Modified Edman	Tarr (1975)
Peptide	100	20	Edman method as modified by Peterson et al. (1972)	Tomita et al. (1978)
Peptide	100–250	21	Edman	Tanaka et al. (1977)
Ferredoxin, Chlorobium thiosulphatophilum	110	22	Edman	Hase et al. (1978)
IgG (heavy chain)	600	20	Dansyl–Edman	Percy and Buchwald (1972)
Peptide	250	20	Dansyl–Edman (modified)	Meagher (1975)
Lysozyme	10	19	Using dabsyl isothiocyanate	Chang (1977)
Ferredoxin, Chromatium vinosum	—	19	Edman	Hase et al. (1977)

It may, in certain circumstances, be used to determine the N-terminal sequence of large peptides, particularly if the residues in this region present difficulties of identification. In essence, after each step of the degradation a small sample of the remaining peptide is withdrawn and its amino acid composition determined. The difference between this and the analysis at the previous step is an indication of which amino acid was removed. An example of the use of this method to the peptide

Cya–His–Thr–Ala–Tyr–Gly–Lys

is given in Table XXVI.

Table XXVI. Subtractive Edman degradation: an example of its application to the heptapeptide Cya–His–Thr–Ala–Tyr–Gly–Lys

Amino acid*	Cya	Thr	Gly	Ala	Tyr	His	Lys
Step 0	1.0	0.8	1.1	1.0	0.9	0.8	1.0
Step 1	0.1	0.8	1.2	1.0	1.0	0.9	1.0
Step 2	—	0.9	1.2	1.0	1.0	0.1	1.0
Step 3	—	0.0	1.1	1.0	1.0	—	1.0
Step 4	—	—	1.0	0.0	1.0	—	1.0
Step 5	—	—	1.0	—	0.1	—	1.0
Step 6	—	—	0.2	—	—	—	1.0

*The composition of the peptide was determined at each step of the Edman degradation by amino acid analysis. The results are expressed as mols of amino acid residue per mol of lysine.

(2) N-Terminal Determination with Dansyl Chloride

The use of 'Dansyl' (5-dimethylamino naphthalene-1-sulphonyl) chloride as a labelling reagent was discovered by accident, by B. S. Hartley, during studies on chymotrypsin. It was found to react specifically with amino groups in this protein, forming derivatives that were stable to subsequent acid hydrolysis, and Hartley considered it would make a suitable N-terminal reagent, as its sensitivity was at least a hundred-fold greater than Sanger's reagent (2,4-dinitrofluorobenzene) (Hartley 1970). The mechanism of the reaction of this reagent with polypeptides is illustrated in Figure XXXIX.

High-voltage electrophoresis was originally used to separate the dansyl amino acids formed after hydrolysis of a dansylated peptide, however the usefulness of this reagent was vastly enhanced by the introduction of thin layer chromatography on polyamide layers for the separation of these

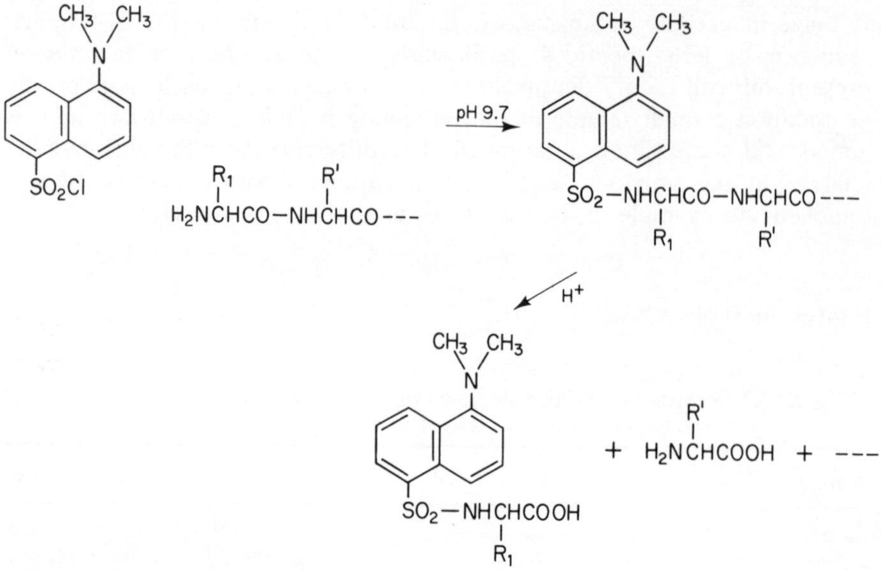

Figure XXXIX Labelling the N-terminal residue of a peptide with dansyl chloride

derivatives. The most useful system is that introduced by Woods and Wang (1967): this is a two-dimensional thin layer chromatography on polyamide layers (Figure XL). Not all the normal dansyl amino acid derivatives are resolved by two chromatographic runs, so a third solvent, run in the same direction as the second, may be employed to resolve the dansyl derivatives of histidine, arginine and cysteic acid. Further, this step improves the separation between Thr/Ser, and Asp/Glu. The replacement of benzene with toluene in the second solvent system, resulting in virtually the same separation, has been suggested, on health considerations (Croft 1972).

The dansylation reaction, followed by subsequent separation of the dansyl-amino acids on polyamide, is an excellent method of N-terminal analysis: however, care has to be exercised to ensure that the dansylated peptide, or protein, has been completely hydrolysed, as this can lead to misinterpretation of the products when identified from polyamide thin layer plates; van Beynum *et al.* (1977) discovered such a mistake in the sequence of alfalfa mosaic virus coat protein. Initially the N-terminal of a tryptic peptide was identified as tyrosine, but it was in fact Ile-Leu. This was because the slowly hydrolysed dipeptide DNS-Ile-Leu appeared on polyamide layers in a similar position to DNS-Tyr. A similar finding was made by Dunkley and Carnegie (1974) in their work on rat myelin basic protein, during which they found that the related dansylated dipeptide DNS-Ile-Val was only slowly hydrolysed by acid. A compromise approach has to be reached, as too long a hydrolysis period is likely to destroy any DNS-Pro, if present.

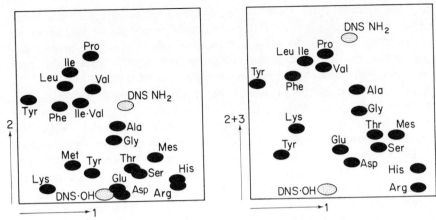

Figure XL   Separation of dansyl amino acids on thin layers of polyamide according to the method of Woods and Wang (1967)

The actual dansylation procedure for peptides is extremely simple: to approximately 100 pmol of peptide is added 100 μl of 0.2 M NaHCO$_3$ followed by 100 μl of 0.25% dansyl chloride in acetone. Reaction takes place at 37°C for 30 min, whereupon the reaction mixture is reduced to dryness. Hydrochloric acid (6 N, 200 μl) is added to the residue and the sample hydrolysed. The lower limit of detection is about 10 pmol. Proteins may be dansylated in 0.2 M NaHCO$_3$ containing 0.1% sodium dodecyl sulphate, after which they are precipitated with acetone, and washed with the same solvent before hydrolysis (Bennett *et al.* 1978). Alternatively proteins may be dansylated in solutions containing 6 M urea, from which they are precipitated with trichloroacetic acid (Mosesson *et al.* 1973).

Sequence Determination by the Dansyl–Edman Method

The straightforward manual Edman degradation is not without its difficulties. One problem is the partial extraction, and consequent loss, of hydrophobic phenylthiocarbamoyl peptides in the washing after cleavage to remove excess phenylisothiocyanate. Gray and Hartley (1963) introduced the procedure that instead of this extraction process, the reaction mixture is dried down and directly treated with acid. Following cleavage all unwanted products, including the thiazolinone, are extracted with n-butyl acetate and discarded. The N-terminal residue at each step of the degradation is identified by dansylation and hydrolysis of a small sample. This procedure also had the added advantage that the DNS-amino acids were more readily identifiable than the PTH-derivatives. This ingenious innovation has been largely responsible for the great upsurge in sequence information that emerged from the world's protein sequence laboratories during the decade 1964–1974. It has tremendous advantages over other methods, in particular it is an extremely simple procedure to implement, so was therefore available to even the smallest laboratory. The author has found that even the most junior technician can readily master this procedure, and with little experience, achieve about four residues per day.

A micro-version of the procedure has been introduced, by which it is possible to sequence as little as 1 nmol of a peptide, and this has been used to sequence part of Met-tRNA synthetase (Bruton and Hartley 1970). The reaction is carried out in micro-tubes (0.2 cm × 8 cm) and after reaction the DNS-amino acids are identified on small sheets of polyamide (7.5 × 7.5 cm). The procedure can also be applied to intact proteins: Weiner *et al.* (1972) have applied the method to proteins eluted from SDS-gels, determining up to 10 residues using only a few nmol of protein, and Percy and Buchwald (1972) have determined 20 residues of the N-terminal sequence of the heavy chain of IgG Sac.

For small peptides, of which the protein chemist might at some stage be encumbered with a large number, a procedure has been devised for rapid sequence determination (Gray and Smith 1970). Essentially, the peptide solution is divided into a number of portions and each one subjected to a successively greater number of cycles. No extraction is performed until the required number of cycles is complete, and all the samples are extracted simultaneously, saving considerable time. Dansylation is then performed, which should clearly indicate the sequence, e.g.

No. of cycles	Peptide resulting	N-terminal residue identified by DNS-Cl
0	Gly–Ala–Ser–Try	DNS-Gly
1	Ala–Ser–Tyr	DNS-Ala
2	Ser–Tyr	DNS-Ser
3	Tyr	DNS-Tyr

One of the most interesting and potentially useful modifications of the Dansyl-Edman degradation has been the enclosed nitrogen chamber version devised by Meagher (1975). This chamber (Figure XLI) is easily constructed in the laboratory and has been found to eliminate problems associated with the oxidation of phenylthiocarbamoyl peptides, and to reduce the loss of protein caused by hydrolysis, as a consequence of moisture in the trifluoroacetic acid used in the cleavage stage. Coupling is performed in aqueous 50% pyridine solution, containing 10^{-5} M dithiothreitol, with the addition of sodium dodecyl sulphate if the protein failed to dissolve well. The method was found to be capable of sequencing up to 12 samples simultaneously and with a speed approaching that of the automatic sequencer.

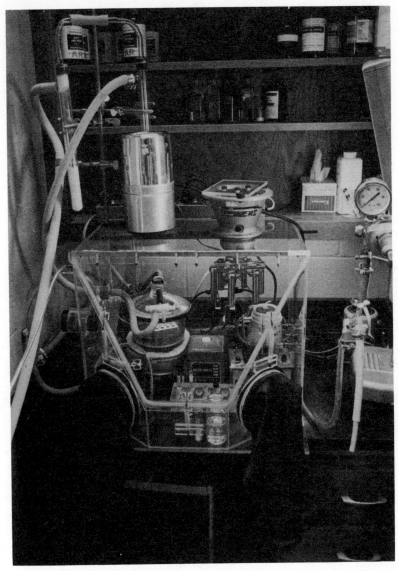

Figure XLI Photograph of the nitrogen chamber used by Meagher to sequence multiple samples of peptides. (Reproduced by courtesy of Dr R. B. Meagher)

(3) Other Useful N-terminal Reagents

Innumerable reagents for N-terminal determination have been proposed, but few have been chosen for practical application to proteins, and only a small number of the most useful will be covered here.

(i) Dabsyl Chloride

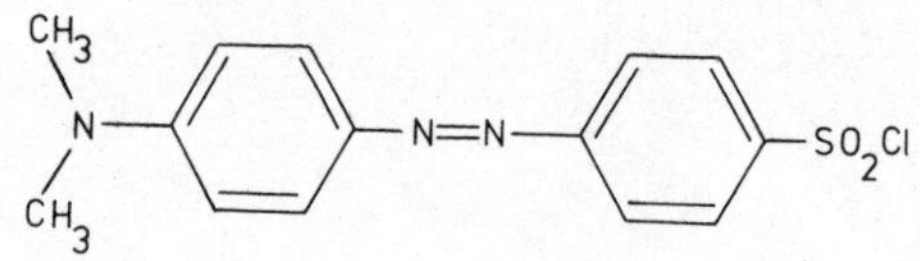

Dabsyl chloride (4-dimethyl aminoazobenzene-4′-sulphonyl chloride) was introduced as a chromophoric label for N-terminal residue determination of peptides and proteins, by Lin and Chang (1975). It has been found to be extensively sensitive, detectable in the 10^{-10} mole range by its intense colour.

(ii) Dabsyl Isothiocyanate

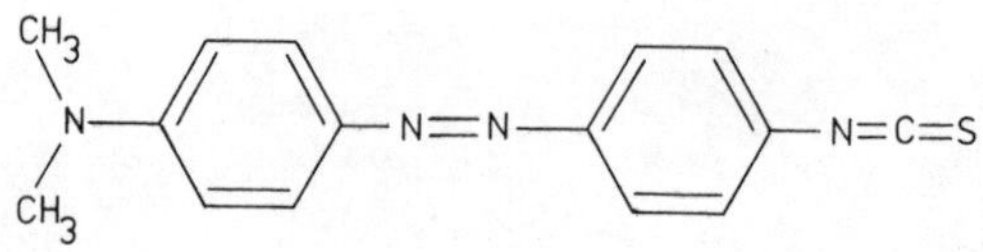

Dabsyl isothiocyanate (4,N,N-dimethyl aminoazobenzene 4′isothiocyanate) was introduced by Chang (1977) as an alternative to phenylisothiocyanate. It has the advantage that the thiazolinones formed on conversion to their thiohydantoin derivatives are readily visualized on polyamide thin layers, after chromatography, due to their intense colour. It was found possible to sequence a hexapeptide completely using only 2 nmol, and it was further used to sequence the first 12 residues of lysozyme employing 10 nmol of protein. It has however, the slight drawback of not reacting as readily with α-amino groups as does phenylisothiocyanate and rather drastic conditions are needed to affect reaction. Chang (1978) has made the interesting observation that this reagent reacts with N-terminal N-monomethyl amino acids directly to form the thiohydantoin ring and he has suggested that this might be the basis of a slightly new degradation procedure.

This reagent has also been attached to porous glass beads and used in a manual solid phase sequencing procedure (Chang *et al.* 1977). With this method it was possible to sequence up to 19 residues using less than 2 nmol of a protein; it was estimated that <0.1 nmol was needed at each step for identification.

(4) Use of Enzymes for N-terminal Determination

The aminopeptidases are a group of intracellular enzymes that hydrolyse amino terminal residues of peptides and proteins. Swine kidney is the major source of these enzymes, the two principal ones being leucine aminopeptidase (LAP) isolated from the supernatant fraction and aminopeptidase M from the microsomal fraction. These two enzymes have many properties in common but differ in their divalent cation requirements and in their specificities. In general they both show a broad specificity, with amino acids having hydrophobic side chains most easily hydrolysed: however, sequences of the type X–Pro are not cleaved by aminopeptidase M (van Beynum *et al*. 1977), e.g.

$$\overrightarrow{Ile}-\overrightarrow{Ala}-\overrightarrow{Leu}-Gln-Pro \ldots \ldots$$

LAP before use requires activation with Mg^{++} and assays are usually performed in tris buffers (0.1 M) at pH 8.6, whereas aminopeptidase M requires no activation and digests are performed in sodium phosphate buffers at pH 7.0. In general, after addition of the enzyme to the substrate, the release of amino acids from the N-terminus is measured kinetically and this gives some indication of the sequence in this region of the protein.

(5) The Determination of Blocked N-terminal Residues

(i) N-terminal Acetyl Groups

Interest in proteins having N-terminal acetyl groups has grown rapidly since Narita (1958) first demonstrated the presence of an N-terminal acetyl group in tobacco mosaic virus protein. Since that time many proteins have been found to have acetylated termini: some of these are tabulated in Table XXVII.

The N-terminal acetyl residue cannot be determined directly on the intact protein. The most precise approach is to isolate the blocked, ninhydrin-negative peptide from an enzymic digest of the protein and to elucidate its structure using mass spectrometry as described in chapter 8. However, chemical methods are quite satisfactory and in general involve digesting the protein with a proteinase of broad specificity, such as pronase from *Streptomyces griseus*. The acetyl-peptide may readily be isolated from the resulting mixture after chromatography on a column of Dowex 50(H^+) (Mok and Waley 1968).

(ii) Pyroglutamyl Residues

Pyrrolidone carboxylic acid (pyroglutamic acid) was found by Wilkinson *et al*. (1966) to be the N-terminal amino acid of the heavy chain of rabbit IgG. Subsequently it has been found to be present in many other proteins and peptides (see Table XXVIII). It is thought that it is formed when the peptide, or protein, is exposed to a slightly acidic environment, by

Table XXVII. Some proteins containing N-acetyl groups

Ac-Gly	Ac-Ser	Ac-Ala	Ac-Thr	Ac-Met	Ac-Trp	Ac-Val
Cytochrome c from: elephant, seal, bonito, horse, carp, dogfish, snapping turtle, kangaroo, zebra, hog, man, camel, ox, chicken, rabbit, *Euglena gracilis*.	Tobacco mosaic virus ribosomal protein, horse liver alcohol dehydrogenase, glyceraldehyde 3-PO_4 dehydrogenase, haemoglobin α-chain of *Catostomus clarkii*, ovine luteinizing hormone, melanocyte stimulating hormone, ox ferritin, histone $f_{2a'}$	Wool keratin, myelin basic protein, carp myogen, cytochrome c from: tomato, leek, spinach, cotton, maidenhair tree, castor, sesame, sunflower, mung bean, rape, buckwheat, cauliflower, pumpkin, wheat germ.	Ox fibrinogen.	α-crystallin.	$β_s$-crystallin.	Chick haemoglobin A1$β$

Table XXVIII. Proteins/peptides having N-terminal pyroglutamic acid

---

Carboxypeptidase inhibitor from potatoes
Cytochrome c' from Alcaligenes
Cytochrome c2 from *Rhodopseudomonas viridis*
Caerulein
Eledoisin
Alytesin
Bombesin
Uperolein
Xenopsin
Ameletin
Neurotensin
Gastrin (pig, man, ovine, ox, cat)
LH- and FSH-releasing hormone
Locust adipokinetic hormone
Blanching hormone
Bovine para kappa casein
Bovine kappa beta 1 casein
Alpha acid glycoprotein from human plasma
Human apolipoprotein Apolp GLN
Human apolipoprotein AII Rhesus monkey

Note: This list excludes the immunoglobulin chains.

cyclization of a precursor glutaminyl residue thus:

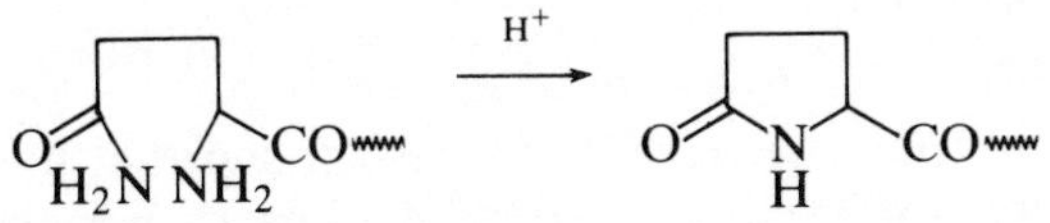

It is unlikely that a glutamyl residue would cyclize under these conditions, as has been proposed by certain workers.

This cyclic structure has posed considerable difficulties during sequence studies, notably in the Edman degradation which is dependent upon an available α-amino group. van Beynum *et al.* (1977) found that they could reduce the extent of cyclization of glutamine residues during the manual Edman degradation by performing the cleavage reaction of the previous residue, with trifluoroacetic acid, at 0°C for 10 min, instead of the usual 40°C and 30 min. O'Donnell and Inglis (1974) have found that the tendency of glutamine to cyclize is enhanced if it is followed in the sequence by a prolyl residue.

There have been many attempts to devise a procedure to chemically open the pyrrolidone ring. The most successful is methanolysis, whereby the peptide is treated at room temperature, with conc. HCl/methanol (1 : 11, v/v) for 2 days. Under these conditions the pyrrolidone ring is opened to give the γ-methyl ester of a glutamyl residue. This procedure was used by van Beynum *et al.* (1977) during an investigation of the structure of Alfalfa mosaic virus coat protein. However, side reactions, particularly hydrolysis of the peptide chain, occur if there are traces of water present (Kawasaki and Itano 1972).

Doolittle (1972) has been successful in isolating an enzyme from certain strains of *Pseudomonas* and *Bacillus subtilis*, that cleave pyroglutamyl residues from polypeptide chains. However, the isolation of these enzymes has posed considerable problems, and Podell and Abraham (1978) have developed a procedure for the release of pyroglutamyl residues from peptides using commercially available pyroglutamate aminopeptidase, isolated from calf liver.

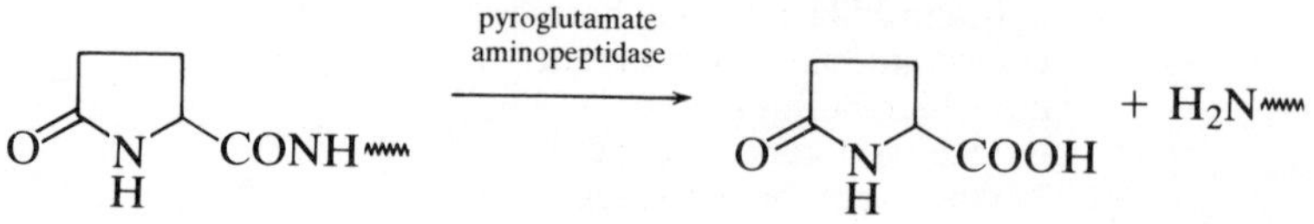

A related cyclization reaction has been found with N-terminal carboxamido-methyl cysteine residues (Gregory and Preston 1977), which condense to form a six-membered lactam ring thus:

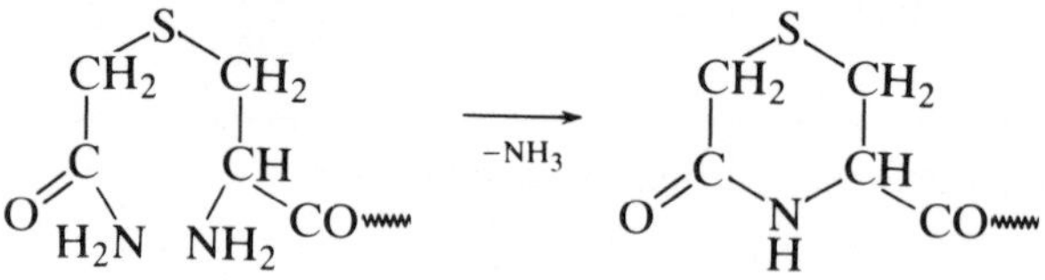

In related studies, Closset *et al.* (1978) have found that during automatic sequencing a very low yield was observed when carboxy-methyl cysteine was the N-terminal residue, and they suggest an analogous cyclization as above.

THE DETERMINATION OF C-TERMINAL RESIDUES AND SEQUENCES

(1) Hydrazinolysis

Hydrazinolysis was introduced by Akabori *et al.* (1952) for the determination of carboxyl-terminal residues in peptides and proteins. The general procedure is to treat the peptide, or protein, with the anhydrous reagent, in a sealed tube, at 100°C, for a period of time between 12 and 24 hours. Hydrazine sulphate is occasionally employed as a catalyst, enabling the reaction to be performed at a lower temperature, although the advantages gained in its use are in doubt. After reaction the excess hydrazine is removed by volatilization *in vacuo*. The reaction is summarized thus:

$$NH_2CHR_1CO \ldots\ldots\ldots NHCHR_nCONH.CHR_{n+1}COOH$$

$$\downarrow N_2H_4$$

$$NH_2CHR_1CONHNH_2, \ldots\ldots NH_2CHR_nCONHNH_2, + NH_2CHR_{n+1}COOH$$

The hydrazine converts all the amino acid residues, except the C-terminal one, into their respective hydrazides. The C-terminal amino acid is then identified by chromatography, after it has been separated from the bulk of the hydrazides. In practice, this is not so easy to accomplish, since the reaction demands anhydrous hydrazine, which is difficult to prepare and dangerous to handle. However in certain circumstances this is an extremely useful procedure. Unfortunatley it is not applicable to C-terminal arginine, asparagine, or glutamine.

(2) Reaction with Carboxypeptidases

The most useful method for the determination of C-terminal residues is undoubtedly the enzymatic procedure. Carboxypeptidase reacts with peptides and proteins to release amino acids that have a free α-carboxyl group. At least four carboxypeptidases have been isolated and described namely, carboxypeptidases A, B, C and Y (Table XXIX). Generally the C-terminal sequence is determined by studying the kinetic release of amino acids from the peptide after the addition of the enzyme. Carboxypeptidase A, which is the best characterised, has been used most.

Table XXIX. Properties of carboxypeptidases (CBPs)

Enzyme	Source	Specificity	Optimum pH
CBP(A)	Pancreas	Removes all amino acids except lysine, arginine and proline	8.0
CBP(B)	Pancreas	Removes only lysine and arginine	8.0
CBP(C)	Citrus leaves	Liberates acidic, basic and neutral amino acids including proline	3.5
CBP(Y)	Baker's yeast	Liberates all C-terminal amino acids including proline	6.0

Usually carboxypeptidases do not react with native proteins, but there are exceptions, for example carboxypeptidase A was found to cause rapid release of C-terminal tyrosine from the lens protein γ-crystallin (Croft 1971; see Figure XLII). No further amino acids were liberated, until the protein had been denatured by performic acid oxidation, whereupon the amino acids released clearly indicated the C-terminal sequence of this protein to be (Figure XLII):

wwwVal–Met–Asp–Phe–Tyr–COOH

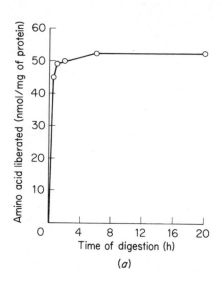

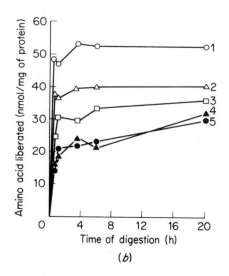

Figure XLII (a) Release of tyrosine from γ-crystallin by carboxypeptidase A (b) Release of amino acids from citraconylated, oxidized γ-crystallin by carboxypeptidase A. Curve 1, tyrosine; curve 2, phenylalanine; curve 3, aspartic acid; curve 4, valine; curve 5, methionine sulphone

Carboxypeptidase C was originally isolated from the peel of citrus fruits by Zuber (1964), who found that it was capable of releasing C-terminal proline, as well as other amino acids. Subsequently it has been prepared from germinated wheat and this preparation has been used in sequence studies on β-purothionin (Mak and Jones 1976). It is interesting that in this study the enzyme reacted with the protein in its native conformation, for as mentioned above this is untypical of the other carboxypeptidases. In related studies, Gregory and Preston (1977) found that whereas carboxypeptidases A and B failed to release amino acids from the C-terminus of urogastrone, the enzyme carboxypeptidase C released them in the expected order. This could possibly be due to partial denaturation of the native substrate in the acidic reaction medium of the carboxypeptidase C digestion.

The enzyme carboxypeptidase Y is isolated from baker's yeast and is sometimes referred to as 'phaseolain'. This carboxypeptidase exhibits a broad specificity towards most C-terminal amino acids, including proline, and may successfully be used in solutions of 6 M urea, as shown by Ponsting1 et al (1978) in their studies of the protein uteroglobin. Lee and Riordan (1978) report that most preparations are contaminated by a specific endopeptidase; however, this activity may be selectively inhibited by the addition of pepstatin A.

(3) C-Terminal Selective Tritium Labelling

Matsuo et al. (1966) introduced an extremely useful and specific method for the C-terminal determination of peptides and proteins. This method has been successfully used by Croft (1968) to determine the C-terminal residues of peptides isolated from the peptide antibiotic viomycin. The reaction may be summarized in the following steps:

(1) the peptide is treated with acetic anhydride to form a C-terminal oxazolone;
(2) base-catalyzed racemization is induced at the C_4 position of the oxazolone ring by pyridine; this is carried out in the presence of T_2O and tritium is therefore incorporated into the oxazolone ring;
(3) the oxazolone ring is opened to regenerate the C-terminal amino acid, that now contains the tritium in the α-position of this residue;
(4) any exchangeable isotope is removed by successive additions of water followed by evaporations;
(5) the peptide is finally hydrolysed and the amino acids separated by chromatography; the radioactive one is identified readily, as the C-terminal residue.

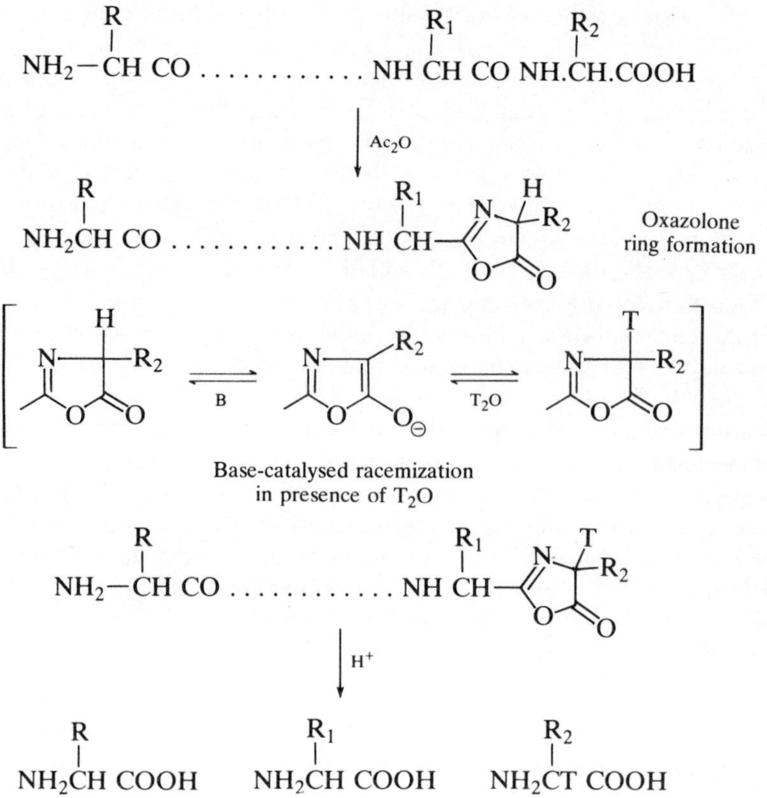

$$\underset{\displaystyle NH_2-\underset{\displaystyle |}{\overset{\displaystyle R}{C}H}\ CO}{} \ldots\ldots\ldots\ldots \underset{\displaystyle NH\ \underset{\displaystyle |}{\overset{\displaystyle R_1}{C}H}\ CO}{}\ \underset{\displaystyle NH.\underset{\displaystyle |}{\overset{\displaystyle R_2}{C}H}.COOH}{}$$

Ac$_2$O

Oxazolone ring formation

Base-catalysed racemization in presence of T$_2$O

H$^+$

$$\underset{NH_2\overset{R}{C}H\ COOH}{} \qquad \underset{NH_2\overset{R_1}{C}H\ COOH}{} \qquad \underset{NH_2\overset{R_2}{C}T\ COOH}{}$$

(4) 'Subtractive' C-terminal Determination

This method, recently introduced by Parham and Loudon (1978), is in principle similar to the old Dakin–West degradation of peptides. The peptide, which is immobilized via its amino terminus to porous glass beads, is treated with di-*p*-nitrophenylphosphoryl azide to convert all carboxylic acid groups to their corresponding acylazides, in one step, thus:

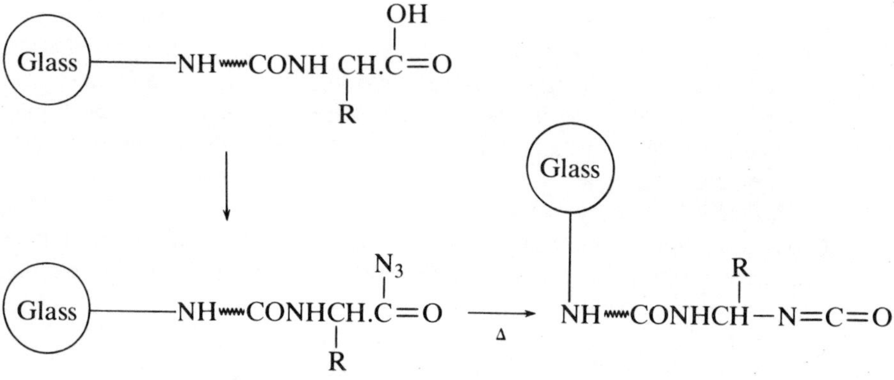

Glass —NHⱲCONH CH.C=O with OH and R

Glass —NHⱲCONHCH.C=O with N$_3$ and R → (Δ) NHⱲCONHCH—N=C=O with R and Glass

Subsequent thermolysis of the acyl azide converts it into an isocyanate. Acid hydrolysis of this product leads to destruction of the C-terminal residue, which may readily be ascertained by the difference in amino acid content before and after reaction. In this respect, it is similar to the subtractive Edman degradation. However, this method is only suitable for small simple peptides, and with peptides of this type Parham and Loudon found that they lost their C-terminal residue in yields greater than 90%.

CHAPTER 8

Applications of Mass Spectrometry to the Sequence Analysis of Peptides and Proteins

In an attempt to devise an automated procedure for the sequence analysis of peptides and proteins, the attention and hope of many chemists, during the last two decades, has been focused on mass spectrometry. Yet despite the early promise of mass spectrometry, that was generated in the early days with studies on small synthetic peptides, peptide antibiotics, and the like, it has not become, contrary to general expectation, the panacea of the protein biochemist. At present the most successful procedure for peptide sequencing by mass spectrometry involves considerable initial chemical modification, including acetylation and permethylation of the peptide, which itself must not have more than, at the most, twelve residues. This derivation is necessary so as to enhance the volatility of the peptide and may further involve other chemical modification reactions. Fragmentation of the modified peptide in the mass spectrometer occurs in a stepwise manner with the loss of a derivatized amino acid residue, one by one, from the C-terminal end of the peptide chain. What occurs is virtually a simultaneous sequential degradation of the peptide from the C-terminus, in contrast to the Edman degradation which begins at the N-terminus of the polypeptide chain and is intermittent. As each amino acid residue derivative is of unique mass (except leucine/isoleucine, see Table XXX) the mass differences of the principal peaks in the resulting spectrum is an indication of the sequence of the peptide. The technique is very sensitive, requiring a similar amount of material as needed by the sequenator, and is reliable. Above all it is a very rapid technique, as up to 10 residues may be determined in one step. Furthermore, it offers the advantage of determining the sequence of peptides that might possess a blocked amino terminus, this being tedious by other means. Mass spectrometry is also capable of accurately analysing and resolving peptide mixtures, so avoiding the time-consuming and wasteful process of peptide isolation and purification.

Table XXX. Integral mass numbers corresponding to derivatives of the amino acids

Amino acid	Structure of derivative	A N-terminal mass	B Mass of residue	C C-terminal mass
Glycine	$\begin{array}{c} CH_3 \\ | \\ -N-CH_2-CO- \end{array}$	114	71	102
Alanine	$\begin{array}{c} CH_3 \\ | \\ -N-CH-CO- \\ | \\ CH_3 \end{array}$	128	85	116
Valine	$\begin{array}{c} CH_3 \\ | \\ -N-CH-CO- \\ | \\ CH(CH_3)_2 \end{array}$	156	113	144
Leucine†	$\begin{array}{c} CH_3 \\ | \\ -N-CH-CO- \\ | \\ CH_2 \\ | \\ CH(CH_3)_2 \end{array}$	170	127	158
Serine	$\begin{array}{c} CH_3 \\ | \\ -N-CH-CO- \\ | \\ CH_2 \\ | \\ O-CH_3 \end{array}$	158	115	146
Threonine	$\begin{array}{c} CH_3 \\ | \\ -N-CH-CO- \\ | \\ CH-O-CH_3 \\ | \\ CH_3 \end{array}$	172	129	160
Aspartic acid	$\begin{array}{c} CH_3 \\ | \\ -N-CH-CO- \\ | \\ CH_2 \\ | \\ CO-OCH_3 \end{array}$	186	143	174
Glutamic acid	$\begin{array}{c} CH_3 \\ | \\ -N-CH-CO- \\ | \\ CH_2 \\ | \\ CH_2 \\ | \\ CO-O-CH_3 \end{array}$	200	157	188

Table XXX. (*continued*)

Amino acid	Structure of derivative	A N-terminal mass	B Mass of residue	C C-terminal mass
Asparagine	CH_3 │ $-N-CH-CO-$ │ CH_2 │ $CO-N(CH_3)_2$	199	156	187
Glutamine	CH_3 │ $-N-CH-CO-$ │ CH_2 │ CH_2 │ $CO-N(CH_3)_2$	213	170	201
Phenyl alanine	CH_3 │ $-N-CH-CO-$ │ CH_2-⬡	204	161	192
Tyrosine	CH_3 │ $-N-CH-CO-$ │ CH_2-⬡$\backslash$ H_3CO	234	191	222
Tryptophan	CH_3 │ $-N-CH-CO-$ │ CH_2- (indole ring) │ CH_3	257	214	245
Lysine	CH_3 │ $-N-CH-CO-$ │ $(CH_2)_4-N(CH_3)$ │ $CO-CH_3$	241	198	229

Table XXX. (*continued*)

Amino acid	Structure of derivative	A N-terminal mass	B Mass of residue	C C-terminal mass
Histidine	CH_3 | $-N-CH-CO-$ | CH_2 | Imidazole$-CH_3$	208	165	196
Ornithine‡	CH_3 | $-N-CH-CO-$ | $(CH_2)_3$ | $N(CH_3)$ | $CO-CH_3$	227	184	215
Proline	$\overset{H_2}{C}$ $CH_2 \quad CH_2$ $-N-CH-CO-$	140	97	128

†Not differentiated from isoleucine.
‡Result of hydrazinolysis of arginine.

HISTORICAL DEVELOPMENT OF THE MASS SPECTROMETRIC
METHOD OF PROTEIN/PEPTIDE SEQUENCING

The mass spectrometer has only during the last two decades become a major weapon in the armoury of the natural product chemist. And the technique has belatedly been applied to proteins and peptides derived thereof. The somewhat slow appreciation of the potential of mass spectrometry in peptide sequencing is undoubtedly due to the well-known reputation that peptides had of being involatile, as well as lacking many of the other aesthetic properties generally possessed by organic substances. However, during investigations on the mass spectrometry of peptide antibiotics and related naturally occurring peptides, the possibilities of this method were first realized.

Early attempts were directed at increasing the volatility of the peptide. Bieman and Vetter (1960) reduced the peptide with LiAlH$_4$ to give a polyaminoalcohol of enhanced volatility, that gave a relatively simple mass spectrum.

$$\text{H(NH CHCO)}_n\text{OH} \xrightarrow{\text{LiAlH}_4} \text{H(NH CH.CH}_2)_n\text{OH}$$

with R below the first structure and R below the second structure.

Electron impact of these derivatives gave principal ions that corresponded to the stepwise loss of amino acid units from the C-terminal end of the molecule and enabled simple peptide sequences to be deduced. However with amino acids with reactive side chains further chemical modification was necessary and the problems encountered led to the abandonment of this approach.

An alternative approach to obtain sufficiently volatile peptide derivatives was to employ the N-acyl peptide ester. Andersson (1958) attempted to do this using the trifluoroacetyl peptide ester, and these studies were extended by Stenhagen (1961). The cleavage patterns obtained with these derivatives were further studied by Prox and Weygand (1967). However, the most successful investigation of these early attempts was by the Russian group, led by Professor M. M. Shemyakin (Shemyakin *et al.* 1966). These investigations showed that it was possible to determine the sequences of peptides containing most of the common amino acids. The volatility of the peptide was enhanced by preparing the tert-butyl esters of the dodecyl peptide derivative. They found that the sequence of the peptide was readily observed from an analysis of the fragmentation pattern, which normally involved rupture of the peptide ester, followed by subsequent loss of each of the amino acid residues thus:

$$\text{R}-\text{CO}\;|\;\text{NH}-\text{CH CO}\;|\;\text{NH CH CO}\;|\;\text{NH CH}-\text{CO}\;|\;\text{OR}$$

with R_1, R_2, R_3 above the respective CH groups.

This technique was found to work on a wide selection of peptides up to a nonapeptide, although this was not seen to be a limiting case.

Although these valiant efforts were seen to offer some hope to the protein chemist, they nevertheless did not bring the technique any nearer to becoming a practical possibility, and it was evident that a completely new approach to the problem was necessary. Fortunately this was on the horizon, and the breakthrough began with the structural work carried out on the peptidolipid fortuitine, isolated from *Mycobacterium fortuitum*. Preliminary chemical investigations on this substance indicated it to be essentially an N-acyl-oligopeptide methyl ester, but a total structure for the molecule could not be proposed. Fortunately, at that moment, M. Barber at the A.E.I. factory in Manchester was looking for high molecular weight compounds to test the performance of his new MS9 mass spectrometer, and it was to him that Professor E. Lederer forwarded a few mg of fortuitine. A mass spectrum was obtained showing two parent peaks at 1331 and 1359, corresponding to $C_{70}H_{125}N_9O_{15}$ and $C_{72}H_{129}N_9O_{15}$ respectively, due to its being a mixture of two homologues containing a C_{20} or a C_{22} fatty acid. Furthermore, it was possible to interpret the spectrum obtained, and the following structure for the peptidolipid was proposed (Barber *et al*. 1965):

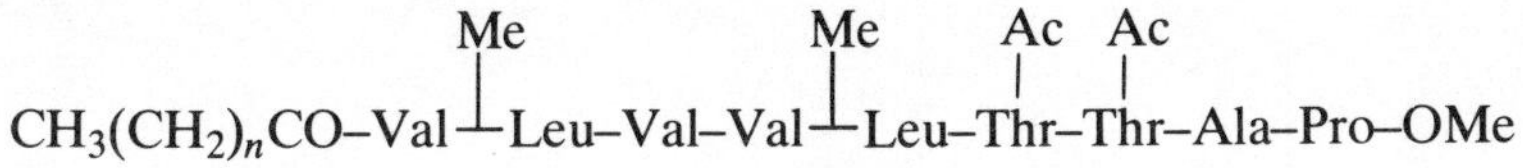

$$\underset{Me}{|} \quad \underset{Me}{|} \quad \underset{Ac}{|} \; \underset{Ac}{|}$$
$$CH_3(CH_2)_nCO-Val\overset{|}{-}Leu-Val-Val\overset{|}{-}Leu-Thr-Thr-Ala-Pro-OMe$$

$$(n = 18, 20)$$

Consideration of the success with fortuitine led workers to suspect that the major limiting factor in the volatility of oligopeptides was the inter-chain hydrogen bonding between the –CO–NH– groups, for fortuitine possessed three tertiary amide bonds. Simultaneous work on the peptide antibiotics reinforced these impressions, as many peptide antibiotics in common with fortuitine contained N-methyl amino acids. These unusual amino acids, as in the peptides enniatin and etamycin, rendered the peptides soluble in organic solvents and enhanced their volatility, so much so that mass spectrometry was feasible. It was therefore reasoned that this increase in volatility was due to a reduction in the hydrogen bonding between the peptide chains, and it was this fact that spurred Das and his colleagues to modify peptides by N-methylation (Das *et al*. 1967). The results were more than encouraging. In addition to giving excellent spectra there was the added bonus that the sequence ions were significantly more intense, making these derivatives suitable for low resolution work. Yet although the N-methylation procedure greatly extended the potential usefulness of mass spectrometry in amino acid sequencing, nevertheless it was not without significant difficulties. In the original procedure the acyl peptide ester was treated with methyl iodide in dimethylformamide and silver oxide.

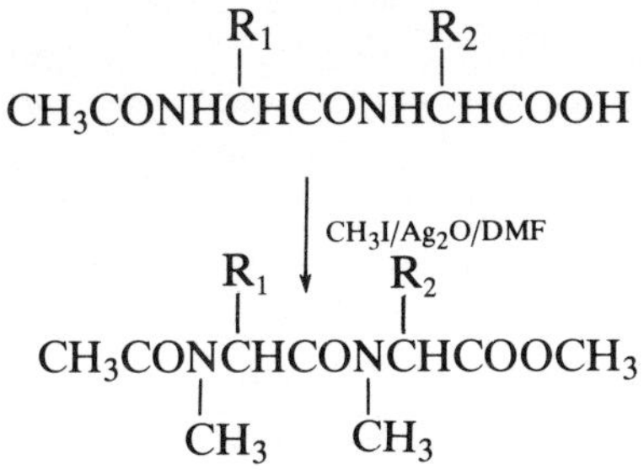

$$\underset{R_1}{|} \qquad \underset{R_2}{|}$$
$$CH_3CONHCHCONHCHCOOH$$

$$\Bigg\downarrow \; CH_3I/Ag_2O/DMF$$

$$\underset{R_1}{|} \qquad \underset{R_2}{|}$$
$$CH_3CONCHCONCHCOOCH_3$$
$$\underset{CH_3}{|} \qquad \underset{CH_3}{|}$$

This treatment resulted in quantitative methylation of the peptide nitrogen atoms and side-chain functionalities, yielding permethyl derivatives of enhanced volatility. Agarwal *et al*. (1969) applied this procedure to gastrin and related peptides and discovered that the permethylation was not straightforward, in particular when the residues were methionine, glutamic acid and aspartic acid. Chain cleavage was found to occur with glutamyl

128

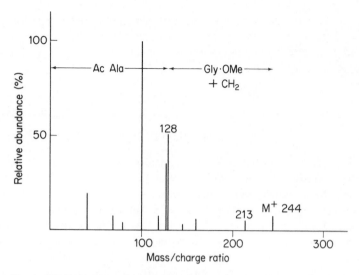

Figure XLIII   The mass spectrum of acetylated and permethylated
Ala–Gly showing C-methylation. M^+, molecular ion

and methionyl residues, and in the case of aspartyl residues even more
complex reactions occurred. At this stage, Thomas (1969) proposed using
the Coggins and Benoiton method, previously employed in the O-per-
methylation of carbohydrates, as a means of permethylating peptides.
This involved treating the peptide in dimethyl formamide with sodium
hydride followed by methyl iodide. However, side reactions were found to
occur. One problem was extensive C-methylation, in particular glycine
residues were prone to this, as may be observed in the mass spectrum of
Ac–Ala–Gly–OH after permethylation using this method (Figure XLIII).

An alternative methylation procedure using methyl iodide and 'dimsyl'
sodium (methyl sulphinyl sodium), as originally described for
carbohydrates by Hakomori (1964), was used by Vilkas and Lederer
(1968). And this procedure was subsequently improved by reducing the
reaction time and limiting the formation of undesirable by-products of
peptides containing the troublesome amino acids, in particular aspartic
acid, glutamic acid and tryptophan (Thomas 1968). It is this method that
has become the basis of the most successful current procedure.

Although most of the usual amino acids were found to give permethyl
derivatives that resulted in successful spectra, the method until recently
was not applicable to several of the common amino acids, including
arginine, histidine, cysteine or its derivatives, and methionine. An attempt
to solve the problem for the sulphur-containing amino acids was made by
Thomas *et al.* (1968), who recommended desulphurization with Raney
nickel whereby the methionine residue was converted to a residue of
α-amino butyric acid. And in related studies quaternary salt formation of

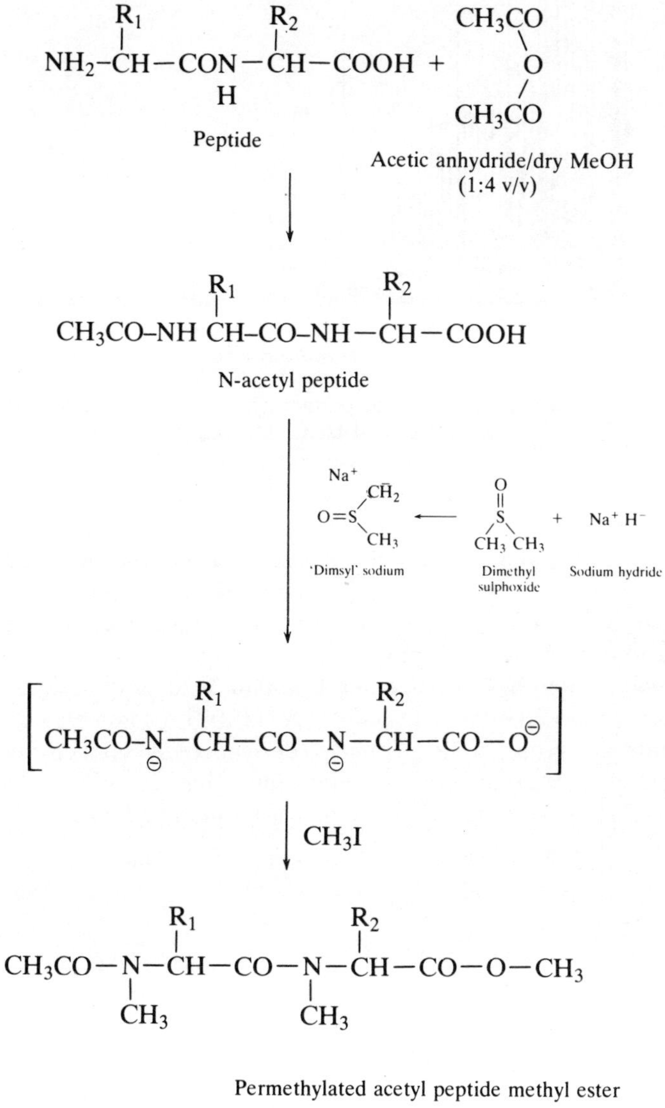

$$\underset{\text{Peptide}}{NH_2\text{--}\underset{\underset{R_1}{|}}{CH}\text{--}CON\text{--}\underset{\underset{R_2}{|}}{CH}\text{--}COOH} \;+\; \underset{CH_3CO}{\overset{CH_3CO}{\diagdown}}O\diagup$$

Acetic anhydride/dry MeOH
(1:4 v/v)

$$CH_3CO\text{--}NH\;\underset{\underset{R_1}{|}}{CH}\text{--}CO\text{--}NH\text{--}\underset{\underset{R_2}{|}}{CH}\text{--}COOH$$

N-acetyl peptide

$$O{=}S\overset{CH_2}{\underset{CH_3}{\diagup}}\quad\longleftarrow\quad \overset{O}{\overset{\|}{S}}\underset{CH_3\;\;CH_3}{}\;+\;Na^+\,H^-$$

Na⁺

'Dimsyl' sodium   Dimethyl sulphoxide   Sodium hydride

$$\left[CH_3CO\text{--}\underset{\ominus}{N}\text{--}\underset{\underset{R_1}{|}}{CH}\text{--}CO\text{--}\underset{\ominus}{N}\text{--}\underset{\underset{R_2}{|}}{CH}\text{--}CO\text{--}O^{\ominus} \right]$$

CH_3I

$$CH_3CO\text{--}\underset{\underset{CH_3}{|}}{N}\text{--}\underset{\underset{R_1}{|}}{CH}\text{--}CO\text{--}\underset{\underset{CH_3}{|}}{N}\text{--}\underset{\underset{R_2}{|}}{CH}\text{--}CO\text{--}O\text{--}CH_3$$

Permethylated acetyl peptide methyl ester

methionine was avoided by its temporary conversion to the sulphoxide, followed by permethylation under normal conditions and reduction back to N-methyl methionine (Roepstorff *et al*. 1970). A more successful approach to the problem was made by Polan *et al*. (1970) who carefully balanced the amount of base used in the deprotonation step of the permethylation procedure with the amount of methyl iodide added. More recently, Jones

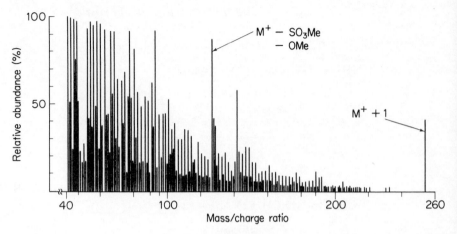

Figure XLIV   The mass spectrum of permethylated and acetylated cysteic acid (by courtesy of Dr G. M. Jones)

(1976) has found that successful mass spectra may be obtained from the fully oxidized derivative of the sulphur-containing amino acid, and Figure XLIV shows the mass spectrum of permethylated cysteic acid, which shows a clear and distinct molecular ion.

Arginine is another problematical amino acid and various strategies have been proposed. Vetter-Diechtl *et al.* (1968) recommended converting the guanidinyl group to a pyrimidyl derivative by condensation with a 1,3-diketone. Alternatively arginine-containing peptides have been derivatized after the arginyl residue had been modified by hydrazinolysis to an ornithyl residue thus:

A different approach to the derivatization of problematical amino acids was made by Morris (1972), who made a study of the rate of permethylation using the Hakomori procedure. It was found that using a shortened reaction time during the permethylation step, quaternization on histidine residues was appreciably slower than the rate of methylation. The success of this method was subsequently extended to include the sulphur-containing amino acids and arginine (Morris *et al.* 1973). This 'short' permethylation procedure is probably the best that is currently available and has been found to give good spectra with most peptides so far encountered. However, recently, Mahajan and Desiderio (1978) have described a new procedure whereby the permethylated peptides are reduced with borane–tetrahydrofuran and the products so formed have been found to be more volatile than the permethylated ones. The sensitivity was also increased, as little as 10 nmol of peptide being sufficient for sequence analysis.

To summarize this section, the following advantages are gained on permethylation of peptides:

(1) there is a decrease in zwitterionic character;
(2) there is an increase in volatility of the peptide due to an inhibition of interchain hydrogen bonding; and
(3) it stabilizes the particular fragmentation pathway, thereby giving maximum sequence information.

OUTLINE OF THE TECHNIQUE

Permethylation is carried out by adding the carbanion solution to the peptide dissolved in dimethyl sulphoxide. Methyl iodide is then carefully added under nitrogen and the reaction allowed to proceed for between 2 and 30 min. The reaction is terminated by the addition of excess distilled H_2O and the permethylated peptide extracted into chloroform. The sample is reduced to dryness and dissolved in a small volume of $CHCl_3$ (approximately 20 μl). By means of a fine capillary tube it is transferred and slowly dried on the quartz tip of the direct insertion probe. The probe is then introduced into the ion source of the mass spectrometer by use of the standard vacuum lock system, and to obtain the spectrum a steady temperature gradient is effected by fractional movement of the probe tip into the ion source. As it moves nearer the electron beam the temperature is slowly increased and the spectrum is continuously monitored until the peptide is seen to volatilize. At this point the temperature gradient is increased rapidly and the spectrum recorded.

Fragmentation of Peptides

Derivatized peptides when subjected to electron impact rupture principally at the amide bonds in two main modes (Shemyakin *et al.* 1966). The principal cleavage is that of the CO–N bond giving rise to the acylium ion, which can then lose the next amino acid residue by loss of carbon monoxide and a neutral imine fragment thus:

$$\overset{b}{} \qquad \overset{a}{}$$

$$\text{wwwCO} - \underset{\underset{CH_3}{|}}{N} - CH - CO \; \vdots \; \underset{\underset{CH_3}{|}}{N} - \underset{\overset{|}{R_2}}{CH} - CO \; \vdots \; O - CH_3$$

with R_1 above the first CH.

Fission at a ($-OCH_3$)

$$\text{wwwCO} - \underset{\underset{CH_3}{|}}{N} - \underset{\overset{|}{R_1}}{CH} - CO - \underset{\underset{CH_3}{|}}{N} - \underset{\overset{|}{R_2}}{CH} - CO^+ \quad \text{Acylium ion}$$

Fission at b $\left(\begin{array}{l} -CO, \\ -N{=}CHR_2 \\ \underset{CH_3}{|} \end{array} \right)$

$$-CO - \underset{\underset{CH_3}{|}}{N} - \underset{\overset{|}{R_1}}{CH} - CO^+$$

Methylation of the amide nitrogen is seen to promote cleavage at each amide linkage with charge retention on the acyl moiety; the resulting 'sequence ions' permit ready assignment of the primary structure of the peptide.

In addition to the peaks corresponding to peptide bond cleavage, the mass spectra of peptides always has additional peaks that correspond to side-chain fragmentation. With leucine, isoleucine and valine, loss of the side chain occurs *via* a McLafferty rearrangement and peaks corresponding to loss of olefinic fragments may be taken as diagnostic of these amino acids (Bricas *et al*. 1965):

Serine residues undergo what is thought to be a thermal reaction with the elimination of CH_3OH by a β-elimination reaction thus:

$$-\underset{\underset{\underset{OCH_3}{|}}{\underset{CH_2}{|}}{N}}-\overset{\overset{CH_3}{|}}{CH}-CO- \quad \xrightarrow[-CH_3OH]{\Delta} \quad -\underset{\underset{CH_2}{\|}}{\overset{\overset{CH_3}{|}}{N}}-C-CO-$$

High or Low Resolution Mass Spectrometry?

Morris *et al.* (1974a) have pointed out that the choice of instrument resolving power is important. Some groups have operated the instrument in the high resolution mode as this may give valuable information as to the chemical composition of the ions in the spectrum. However these authors point out and further demonstrate in their study of the enzyme ribitol dehydrogenase that high resolution mass spectrometry is not essential in the analysis of peptides (Morris *et al.* 1974b).

Examples of Its Application to Peptides

The mass spectrum of glutathione is illustrated in Figure XLV. This tripeptide (γ-Glu–Cys–Gly), which contains cysteine, was derivatized by a 'short' permethylation reaction. The spectrum shows the molecular ion at m/e 433 (0.3%), the loss of CH_3O (m/e 31) from the M^+ to m/e 402 (0.4%), loss of the glycyl residue minus H at m/e 330 (1.2%), loss of CO

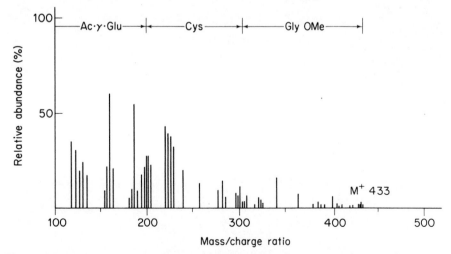

Figure XLV   The mass spectrum of acetylated and permethylate glutathione (by courtesy by Dr G. M. Jones)

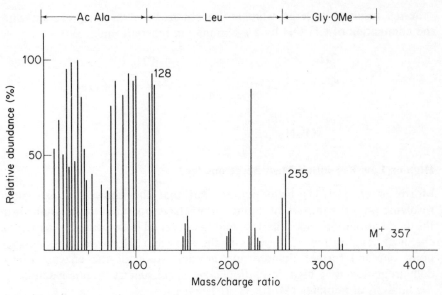

Figure XLVI The mass spectrum of permethylated Ac–Ala–Leu–Gly (by courtesy of Dr G. M. Jones)

(m/e 28) to m/e 302 (0.8%), and loss of the cysteinyl residue to m/e 200 (18%), which is the mass of the N-terminal residue. A further example is of the tripeptide Ala–Leu-Gly (Figure XLVI).

One of the most important advantages of mass spectrometric analysis is that of mixture analysis, as this may reduce the laborious and tedious

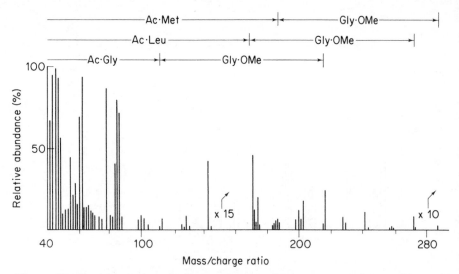

Figure XLVII The mass spectra of a mixture of three dipeptides after acetylation and permethylation (by courtesy of Dr G. M. Jones)

isolation procedures, that have so far been the rate-determining step. Mixture analysis may be accomplished by fractionation, within the mass spectrometer itself, by using a sample temperature gradient. The mass spectrum of a mixture of three simple dipeptides, namely Met–Gly, Leu–Gly, and Gly–Gly, is shown in Figure XLVII, to illustrate the technique.

Application to Proteins

The first published results of the application of mass spectrometry to peptides isolated from proteins was by Geddes *et al.* (1969), who determined the amino acid sequence of a peptide derived from silk fibroin, and shortly after this Agarwal *et al.* (1969) determined the structure of the peptide hormone gastrin. However, a more convincing demonstration of the potential of the mass spectrometric method was the analysis of a mixture of two octadecapeptides from swine immunoglobulin λ-chains (Franek *et al.* 1969). In this study it was possible to determine the sequence of the first 10 residues of the peptides as well as being able to locate the position of microheterogeneity. This early work on peptide mixtures was advanced by the studies of Morris *et al.* (1971), who modified the existing methods to be applicable to micro-quantities of peptide material. Furthermore Morris *et al.* (1974a) argued that high resolution mass spectrometry, involving the determination of the elemental composition of the significant ions in the spectrum, was unnecessary, and that low resolution mass spectrometry could cope adequately with complex peptide mixtures.

Some of the proteins that have been sequenced partially, or completely, by mass spectrometry are given in Table XXXI.

Table XXXI. Some applications of mass spectrometry to proteins

Protein/peptide	Reference
Silk fibroin	Geddes *et al.* (1969)
Feline gastrin	Agarwal *et al.* (1969)
Scotophobin	Desidero *et al.* (1971)
α-Lactalbumin	Bacon and Graham (1972)
Aspartate aminotransferase	Ovchinnikov *et al.* (1973)
Triose phosphate isomerase	Priddle and Offord (1974)
Ostrich cytochrome c	Howard *et al.* (1974)
Dihydrofolate reductase	Morris *et al.* (1974a)
Ribitol dehydrogenase	Morris *et al.* (1974b)
Rabbit myosin (alkali lightchain)	Frank and Weeds (1974)
Tobacco mosaic virus protein	Rees *et al.* (1974)
Prothrombin	Morris (1975)
Concanavalin A	Jones (1976)
Stellacyanin	Bergman *et al.* (1977)
Antifreeze glycoprotein	Morris *et al.* (1978)

Other workers have put forward the opinion that the methods described above are not entirely satisfactory for the resolution of peptides in the complex mixtures derived from the enzymic hydrolysis of proteins, and that alternative methods are needed to prepare peptide derivatives of greater volatility that possibly could be separated by gas chromatography followed by the 'on-line' mass spectrometric characterization of the mixture in a single step. The first attempt to solve the many technical problems was made by Calam (1972). Subsequently work involved the use of the N-trifluoroacetyl, and pentafluoropropionyl peptide methyl esters, but Nau (1974) reported that the O-trimethylsilylpolyamino alcohol obtained by LiAlD$_4$ reduction and trimethylsilylation of the N-acetyl oligopeptide methyl ester was more volatile, with an easily interpretable mass spectrum. Even more volatile derivatives were made by the LiAlD$_4$ reduction and O-trimethylsilylation of the corresponding perfluoroacylated oligopeptide methyl ester and it was found that all the amino acid residues including Arg, His, Trp, Gln and Asn could be derivatized by this technique without modification. In a further study the sequence of a 39 residue peptide was entirely deduced (Nau and Bieman 1976) by mass spectrometry. The peptide, a carboxypeptidase inhibitor from potatoes, was fragmented and the peptides esterified, trifluoracetylated, deuteroalkylated by reduction with LiAlD$_4$ and O-trimethylsilyated. The resulting mixture was separated by GLC and analysed on a mass spectrometer linked up to a computer.

Glycoproteins

The mass spectrometric method has recently been applied to the antifreeze glycoproteins from the blood of the Antarctic fish *Trematomus borchgrevinki* (Morris *et al.* 1978); and the Cu-containing glycoprotein stellacyanin from the Lacquer tree (Bergman *et al.* 1977). In this former study it was found that during the permethylation reaction the 'dimsyl' carbanion abstracts the proton from the α-carbon atom, leading to β-elimination of the carbohydrate moiety thus:

Proteins with Blocked N-terminal Residues

The mass spectra methods differs from the classical approach of sequence determination in the very important fact that it does not demand that the

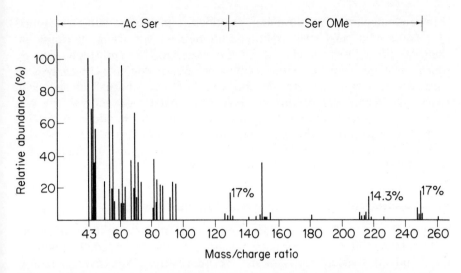

Figure XLVIII   The mass spectrum of permethylated Ac–Ser–Ser isolated from the N-terminus of ox ferritin (by courtesy of Dr I. Al-Hassan)

amino terminus be free. The N-terminus of ox ferritin was shown to be N-acetyl serine by permethylation of a blocked tryptic peptide (Al-Hassan 1976) (see Figure XLVIII), and a similar approach was made by Joassin and Gerday (1977) who deduced that the N-terminus of the major parvalbumin of whiting (*Gadus Merlangus*) was N-acetyl alanine. The strength of the information obtained by mass spectrometric analysis is far greater than that if a chemical method such as hydrazinolysis were employed, though the amount needed for each analysis is about comparable. Auffret *et al.* (1978) showed that the N-terminus of alcohol dehydrogenase from *Drosophila melanogaster* N-11 was N-acetyl serine using 50 nmol of peptide, which was considered to be near the practical limit of the technique.

Recently Pettigrew and Smith (1977) have found that certain proteins have N-terminal α-N-methyl amino acids. The biological significance of these findings is as yet unknown, but these workers have postulated that other proteins might also contain this type of modification. Almost simultaneously it was discovered that the N-terminus of the protein initiation factor IF-3 from *E. coli* was N-methyl methionine (Braver and Wittmann-Liebold 1977). This was determined by mass spectrometric analysis of the N-terminal PTH-derivative after one step of the Edman degradation.

Unusual Amino Acid Residues

Convincing evidence of the power and potential usefulness of the mass spectrometer in sequence analysis has come from studies on the vitamin

K-dependent part of prothrombin. Peptides from this protein were found to possess abnormal electrophoretic mobilities, yet using conventional chemical procedures no clue as to the nature of the 'prosthetic group', which must be acidic in nature, could be found. Amino acid analysis in particular did not indicate the presence of any unusual amino acid. Eventually the nature of this anomaly was solved independently by two groups of workers both employing mass spectrometry. Stenflo *et al.* (1974) studied pure peptides, whereas Magnusson *et al.* (1974) worked on peptide mixtures, however both groups identified the unusual amino acid γ-carboxyglutamic acid in the anomalous peptides.

Mass Spectrometry in Conjunction with Chemical Methods

Burgus *et al.* (1973) have identified some of the residues of ovine hypothalamic luteinizing-hormone-releasing factor by a combination of GLC and mass spectrometry of the PTH-derivatives. Moreover, if methyl isothiocyanate is used in place of the phenyl derivative the thiourea that results when the peptide is reacted has been found to rearrange and cleave to the thiohydantoin in the mass spectrometer ion source (Fairwell *et al.* 1973), and this has been proposed as the basis of a mass spectrometric–Edman method. Alternatively the PTH-amino acid obtained from the automatic sequenator may be identified directly by mass spectrometry. This may also permit the quantitation of these derivatives by concurrent analysis of deuterated analogues (Tschesche *et al.* 1972).

THE USE OF NEW IONIZATION TECHNIQUES

In recent years new ionization techniques have been developed that have prompted workers to investigate their applicability to the sequence analysis of peptides.

(i) Chemical Ionization

Generally chemical ionization gives rise to more complex spectra than does electron impact, however the spectra usually exhibit highly abundant molecular-weight-determining ions, as is to be expected from the 'milder' ionization processes involved. Usually methane reagent gas is employed, though Morris *et al.* (1978) have recently employed ammonia reagent gas in a study of the structure of the glycoproteins of antarctic fish.

Krutzsch and Kindt (1979) have studied the chemical ionization spectra of 40 dipeptides as their trimethylsilylated derivatives and have reported this technique to be at least ten times as sensitive as electron ionization methods. Two intense ions were found in each spectrum, namely $[M + 1]^+$ and $[M - 15]^+$ that allowed unambiguous assignments of the dipeptide molecular weights, and the principal ion in each spectrum was that derived from the N-terminal fragment $[Me_3Si\text{-}NH = CHR_1]^+$, formed by β-cleavage of the central CH–CO bond. Further it was suggested that this method could be used for the identification of dipeptides obtained from a protein after digestion with the enzyme dipeptidyl amino peptidase.

Chemical ionization therefore is a suitable method to supplement, rather than replace, electron ionization methods. However, at present this method has had little application to peptides, which probably reflects the lack of availability of suitable instruments to biochemists.

(ii) Photoionization

The technique of photoionization, as applied to peptides, has been compared to that of electron impact by Orlov *et al.* (1972). In this study it was found that photoionization gave rise to a simpler spectrum, the number of peaks observed, particularly at $m/e < 250$, being considerably reduced. Peptides were observed to fragment in a similar manner to that of electron ionization, however the lack of some sequence ions tended to result in ambiguities and an inability to sequence some regions of the peptide chain.

(iii) Field Desorption

Winkler and Beckley (1972) have applied field desorption mass spectrometry to a few small peptides and the results have been very encouraging in that strong molecular ions have been observed on underivatized peptides, some containing arginine, which would not have been volatile under normal conditions. The fact that no prior treatment of the peptide is needed to increase its volatility enabled smaller quantities of peptide to be used, and the sensitivity of the method is estimated to be approximately 10^{-8} g. However, field desorption studies provide little sequence information for underivatized peptides and the principal use of this technique is in providing a criterion of purity and molecular weight of a peptide. Morris *et al.* (1978) used this technique to determine the nominal mass of an acetyl permethyl derivative of a glycopeptide from the blood of an Antarctic fish. Approximate mass measurement was made on the major ion present which was found to be 1324, which corresponded to a peptide containing 14 amino acid residues.

PROSPECTS

The unique usefulness of mass spectrometry has been clearly shown in work on the identification of unusual amino acids in peptides. The technique is entirely suited to this, as in most instances only minute amounts of material are available and it is difficult to envisage these problems being successfully solved in any other way. The discovery of the amino acid γ-carboxyglutamic acid is such an example. It is clear that with the technique of mass spectrometry, as applied to proteins, so well developed, more of these unusual amino acids will be found probably having been overlooked in previous investigations.

It is clear that mass spectrometry is a valuable tool for the protein biochemist. It is nevertheless an expensive instrument, both to purchase and to maintain, and is obviously not going to be available to most investigators. The best that one may hope for, is to have some form of limited access to a mass spectrometer, though this itself is beset with difficulties. In particular such machines are usually in great demand by chemists, and any particular machine might routinely process 2000–3000 samples per year. For the organic chemist who can usually boost the abundance of high molecular weight ions in the spectrum by increasing the sample size over residual background, this offers no problem—however, this is not possible for the protein biochemist, who usually has only a very small sample available. The upshot is that under these circumstances the detection sensitivity and thus the length of peptide from which sequence information may be determined is about 6–7 residues. Considering that the amount of material required is around 10–30 nmol the technique offers little when compared to established 'wet' chemical methods such as the dansyl–Edman procedure. In conclusion therefore, if a suitable mass spectrometer is not available largely for peptide work, the classical methods are to be preferred, unless the presence of an unusual amino acid is suspected.

References

Agarwal, K. L., Kenner, G. W., and Sheppard, R. C. (1969) Feline gastrin. An example of peptide sequence analysis by mass spectrometry. *J. Amer. Chem. Soc.*, **91**, 3096–3097.

Akabori, S., Ohno, K., and Narita, K. (1952) On the hydrazinolysis of proteins and peptides: A method for the characterization of carboxyl-terminal amino acids in proteins. *Bull. Chem. Soc. Japan*, **25**, 214–218.

Al-Hassan, I. A. A. (1976) Studies on bovine spleen ferritin. *Ph.D. Thesis*, University of Salford.

Andersson, C. O. (1958) Mass spectrometric studies on amino acid and peptide derivatives. *Acta Chem. Scand.*, **12**, 1353.

Atherton, E., Bridgen, J., and Sheppard, R. C. (1976) A polyamide support for solid-phase sequencing. *FEBS Letters*, **64**, 173–175.

Auffret, A. D., Williams, D. H., and Thatcher, D. R. (1978) Identification of the blocked N-terminus of an alcohol dehydrogenase from *Drosophila melanogaster* M-11. *FEBS Letters*, **90**, 324–326.

Bacon, J. R. and Graham, G. N. (1972) Sequence studies on the milk protein α-lactalbumin by using mass spectrometry. *Biochem. J.*, **127**, 76P–77P.

Barber, M., Jolles, P., Vilkas, E., and Lederer, E. (1965) Determination of amino acid sequences in oligopeptides by mass spectrometry. I. The structure of fortuitine, an acyl-nonapeptide methyl ester. *Biochem. Biophys. Res. Commun.*, **18**, 469–473.

Begg, G. S. and Morgan, F. J. (1976) A non-volatile buffer with improved performance in automated protein sequencing. *FEBS Letter*, **66**, 243–245.

Begg, G. S., Pepper, D. S., Chesterman, C. N., and Morgan, F. J. (1978) Complete covalent structure of human β-thromboglobulin. *Biochemistry*, **17**, 1739–1744.

Bennett, C. D., Rodkey, J. A., Sondey, J. M. and Hirschmann, R. (1978) Dihydrofolate reductase: the amino acid sequence of the enzyme from a methotrexate-resistant mutant of *Escherichia coli*. *Biochemistry*, **17**, 1328–1337.

Berg, A. van Den, van Den Hende-Timmer, L., and Beintema, J. J. (1976) Isolation, properties and primary structure of coypu and chinchilla pancreatic ribonuclease. *Biochim. Biophys. Acta*, **453**, 400–409.

Berg, A. van Den, van Den Hende-Timmer, L., Hofsteenge, J., Gaastra, W., and Beintema, J. J. (1977) Guinea-pig pancreatic ribonucleases. Isolation, properties, primary structure and glycosidation. *Eur. J. Biochem.*, **75**, 91–100.

Bergman, C., Gandvik, E., Nyman, P. O., and Strid, L. (1977) The amino acid sequence of stellacyanin from the Lacquer tree. *Biochem. Biophys. Res. Commun.*, **77**, 1052–1059.

Beynum, G. M. A. van, DeGraaf, J. M., Castel, A., Kraal, B., and Bosch, L. (1977) Structural studies on the coat protein of Alfalfa mosaic virus—the complete primary structure. *Eur. J. Biochem.*, **72**, 63–78.

Beyreuther, K., Adler, K., Fanning, E., Murray, C., Klemm, A., and Geisler, N. (1975) Amino acid Sequence of *lac* repressor from *Escherichia coli. Eur. J. Biochem.*, **59**, 491–509.

Beyreuther, K., Bohmer, H., and Dimroth, P. (1978) Amino acid sequence of citrate-lyase acyl-carrier protein from *Klebsiella aerogenes. Eur. J. Biochem.*, **87**, 101–110.

Beyreuther, K., Raufuss, H., Schrecker, O., and Hengstenberg, W. (1977) The phosphoenolpyruvate-dependent phosphotransferase system of *Staphylococcus aureus. Eur. J. Biochem.*, **75**, 275–286.

Bhown, A. S., Mole, J. E., and Bennett, J. C. (1977) Primary structure of human J chain: isolation and characterization of tryptic and chymotryptic peptides of human J chain. *Biochemistry*, **16**, 3501–3507.

Biemann, K. and Vetter, W. (1960) Separation of peptide derivatives by gas chromatography combined with the mass spectrometric determination of the amino acid sequence. *Biochem. Biophys. Res. Commun.*, **3**, 578–584.

Braunitzer, G., Schrank, B., Petersen, S., and Petersen U. (1973) Automatic sequencing of insulin. *Hoppe-Seyler's Z. Physiol. Chem.*, **354**, 1563–1566.

Braunitzer, G., Schrank, B., and Ruhfus, A. (1971) On the complete automatic sequence analysis of peptides using quadrol. *Hoppe-Seyler's Z. Physiol. Chem.*, **352**, 1730–1732.

Braver, D. and Wittmann-Liebold, B. (1977) The primary structure of the initiation factor IF-3 from *E. coli. FEBS Letters*, **79**, 269–275.

Bricas, E., Van-Heijenoort, J., Barber, M., Wolstenholme, W. A., Das, B. C., and Lederer, E. (1965) Determination of amino acid sequences in oligopeptides by mass spectrometry. IV. Synthetic N-acyl oligopeptide methyl esters. *Biochemistry*, **4**, 2254–2260.

Brosius, J. and Arfsten, U. (1978) Primary structure of protein L19 from the large subunit of *Escherichia coli* ribosomes. *Biochemistry*, **17**, 508–516.

Bruton, C. J. and Hartley, B. S. (1970) Chemical studies on methionyl-tRNA synthetase from *Escherichia coli. J. Mol. Biol.*, **52**, 165–178.

Burgus, R., Ling, N., Butcher, M., and Guillemin, R. (1973) Primary structure of somatostatin, a hypothalamic peptide that inhibits the secretion of pituitary growth hormone. *Proc. Natn. Acad. Sci. U.S.A.*, **70**, 684–688.

Butkowski, R. J., Elion, J., Downing, M. R., and Mann, K. G. (1977) Primary structure of human prethrombin 2 and α-thrombin *J. Biol. Chem.*, **252**, 4942–4957.

Butler, W. T., Finch, J. E., and Miller, E. J. (1977) Covalent structure of cartilage collagen. Amino acid sequence of residues 363–551 of bovine α1(II) chains. *Biochemistry*, **16**, 4981–4990.

Calam, D. H. (1972) Gas chromatography of permethylated peptides. *J. Chrom.*, **70**, 146–150.

Carnegie, P. R. (1969) Digestion of an Arg–Pro bond by trypsin in the encephalitogenic basic protein of human myelin. *Nature*, **223**, 958–959.

Carpenter, F. H. and Shiigi, S. M. (1974) Cyanogen bromide treatment of N-acetylmethionyl residues without cleavage. *Biochemistry*, **13**, 5159–5164.

Cavadore, J.-C., Derancourt, J., and Previero, A. (1976) N-aminoethyl poly-acrylamide as support for solid-phase sequencing of proteins. *FEBS Letters*, **66**, 155–157.

Chang, J. Y. (1977) High-sensitivity sequence analysis of peptides and proteins by 4-NN-dimethylaminoazobenzene-4'-isothiocyanate. *Biochem. J.*, **163**, 517–520.

Chang, J. Y. (1978) A novel Edman-type degradation: direct formation of the thiohydantoin ring in alkaline solution by reaction of Edman-type reagents with N-monomethyl amino acids. *FEBS Letters*, **91**, 63–68.

Chang, J. Y., Creaser, E. H., and Hughes, G. J. (1977) A new approach for the solid phase sequence determination of proteins. *FEBS Letters*, **78**, 147–150.

Chauvet, M. T., Codogno, P., Chauvet, J., and Acher, R. (1977) Phylogeny of the neurophysins: complete amino acid sequence of horse MSEL-neurophysin. *FEBS Letters*, **80**, 374–376.

Chauvet, M. T., Codogno, P., Chauvet, J., and Acher, R. (1978) Phylogeny of neurophysins — complete amino acid sequence of whale (*Balaenoptera physalus*) MSEL-neurophysin. *FEBS Letters*, **88**, 91–93.

Chen, B. L., Chiu, Y. H., Humphrey, R. L., and Poljak, R. J. (1978) Amino acid sequence of the human myeloma lambda chain WIN. *Biochim. Biophys. Acta.*, **537**, 9–21.

Chen, R. (1977) Sequence determination of protein S9 from the *Escherichia coli* ribosome. *Hoppe-Seyler's Z. Physiol. Chem.*, **358**, 1415–1430.

Chen, R., Mende, L., and Arfsten, U. (1975) The primary structure of protein L27 from the peptidyl-tRNA binding site of *Escherichia coli* ribosomes. *FEBS Letters*, **59**, 96–99.

Closset, J., Maghuin-Rogister, G., Hennen, G., and Strosberg, A. D. (1978) Porcine follitropin—the amino acid sequence of the β-subunit. *Eur. J. Biochem.*, **86**, 115–120.

Croft, L. R. (1968) The structure of viomycin. *Ph.D. Thesis*, University of Nottingham.

Croft, L. R. (1971) C-terminal amino acid sequence of bovine γ-Crystallin. *Biochem. J.*, **121**, 557–559.

Croft, L. R. (1972a) The amino acid sequence of γ-Crystallin from calf lens. *Chem. Commun.*, **1972**, 437–438.

Croft, L. R. (1972b) The amino acid sequence of γ-Crystallin (fraction II) from calf lens. *Biochem J.*, **128**, 961–970.

Das, B. C., Gero, S. D., and Lederer, E. (1967) N-methylation of N-acyl oligopeptides. *Biochem. Biophys. Res. Commun.*, **29**, 211–215.

DeLange, R. J., Chang, J. Y., Shaper, J. H., and Glazer, A. N. (1976) Amino acid sequence of flagellin of *Bacillus subtilis* 168. *J. Biol. Chem.*, **251**, 705–711.

Deselnicu, M., Lange, P. M., and Heidemann, E. (1973) Studies on the cleavage of the α2 chain of collagen with hydroxylamine. *Hoppe-Seyler's Z. Physiol. Chem.*, **354**, 105–116.

Desidero, D. M., Ungar, G. and White, P. A. (1971) The use of mass spectrometry in the structural elucidation of Scotophobin — a specific behaviour-inducing brain peptide. *Chem. Commun.*, 432–433.

Dijk, H. van, Sloots, B., van den Berg, A., Gaastra, W., and Beintema, J. J. (1976) The primary structure of muskrat pancreatic ribonuclease. *Int. J. Peptide Protein Res.*, **8**, 305–316.

Dixit, S. N., Seyer, J. M., and Kang, A. H. (1977a) Covalent structure of collagen: amino acid sequence of chymotryptic peptides from the carboxyl-terminal region of α2-CB3 of chick-skin collagen. *Eur. J. Biochem.*, **81**, 599–607.

Dixit, S. N., Seyer, J. M., and Kang, A. H. (1977b) Covalent structure of collagen: isolation of chymotryptic peptides and amino acid sequence of the amino-terminal region of α2-CB3 from chick-skin collagen. *Eur. J. Biochem.*, **73**, 213–221.

Dixon, H. B. F. and Perham, R. N. (1968) Reversible blocking of amino groups with citraconic anhydride. *Biochem. J.*, **109**, 312–314.

Doolittle, R. F. (1972) Terminal pyrrolidone carboxylic acid: cleavage with enzymes, in *Methods in Enzymology*, Vol. **XXVB**, 231–244, Eds C. H. W. Hirs and S. N. Timasheff (Academic Press, London).

Dunkley, P. R. and Carnegie, P. R. (1974) Amino acid sequence of the smaller basic protein from rat brain myelin. *Biochem. J.*, **141**, 243–255.

Dwulet, F. E. and Gurd, F. R. N. (1976) A comparison of sulfonated phenylisothiocyanates for reducing losses of lysine-containing peptides during automated sequencing. *Anal. Biochem.*, **76**, 530–538.

Edman, P. (1950) Method for the determination of the amino acid sequence in peptides. *Acta Chem. Scand.*, **4**, 283–293.

Edman, P. and Begg, G. (1967) A protein sequenator. *Eur. J. Biochem.*, **1**, 80–91.

Eerd, J. P. van and Takahashi, K. (1976) Determination of the complete amino acid sequence of bovine cardiac troponin-C. *Biochemistry*, **15**, 1171–1180.

Elliott, D. F. (1952) A search for specific chemical methods for fission of peptide bonds I. *Biochem. J.*, **50**, 542–550.

Emmens, M., Welling, G. W., and Beintema, J. J. (1976) The amino acid sequence of pike-whale (Lesser Rorqual) pancreatic ribonuclease. *Biochem. J.*, **157**, 317–323.

Enfield, D. L., Ericsson, L. H., Blum, H. E., Fischer, E. H., and Neurath, H. (1975) Amino acid sequence of parvalbumin from rabbit skeletal muscle. *Proc. Natn. Acad. Sci. U.S.A.*, **72**, 1309–1313.

Evenberg, A., Meyer, H., Gaastra, W., Verheij, H. M., and deHaas, G. H. (1977) Amino acid sequence of phospholipase A_2 from horse pancreas. *J. Biol. Chem.*, **252**, 1189–1196.

Fairwell, T., Ellis, S., and Lovins, R. E. (1973) Quantitative protein sequencing using mass spectrometry: thermally induced formation of thiohydantoin amino acid derivatives from N-methyl- and N-phenylthiourea amino acids and peptides in the mass spectrometer. *Anal. Biochem.*, **53**, 115–123.

Ferrell, R. E., Stroup, S. K., Tanis, R. J., and Tashian, R. E. (1978) Amino acid sequence of rabbit carbonic anhydrase II. *Biochim. Biophys. Acta*, **533**, 1–11.

Fieser, L. F. and Fieser, M. (1967) *Reagents for Organic Synthesis* (John Wiley and Sons Ltd, New York).

Fleer, E. A. M., Verheij, H. M., and deHaas, G. H. (1978) The primary structure of bovine pancreatic phospholipase A_2. *Eur. J. Biochem.*, **82**, 261–269.

Fontana, A. (1972) Modification of tryptophan with BNPS-skatole (2-(2-nitrophenylsulfenyl)-3-methyl-3-bromoindolenine) in *Methods in Enzymology*, Vol. **XXV**, 419–423 Eds C. H. W. Hirs and S. N. Timasheff (Academic Press, London).

Foriers, A., DeNeve, R., Kanarek, L., and Strosberg, A. D. (1978) Common ancestor for concanavalin A and lectin from lentil? *Proc. Natn. Acad. Sci. U.S.A.*, **75**, 1136–1139.

Foster, J. A., Bruenger, E., Hu, C. L., Albertson, K., and Franzblau, C. (1973) A new technique for automated sequencing of non-polar peptides. *Biochem. Biophys. Res. Commun.*, **53**, 70–74

Fowler, A. V. and Zabin, I. (1978) Amino acid sequence of β-galactosidase XI. Peptide ordering procedures and the complete sequence. *J. Biol. Chem.*, **253**, 5521–5525.

Fraenkel-Conrat, H., Harris, J. I., and Levy, A. L. (1955) Recent developments in techniques for terminal and sequence studies in peptides and proteins. *Methods Biochem. Anal.*, **2**, 393–425.

Franek, F., Keil, B., Thomas, D. W., and Lederer, E. (1969) Chemical and mass spectral sequence of a peptide from the variable part of normal immunoglobulin lambda chains. *FEBS Letters*, **2**, 309–312.

Frank, G., Sidler, W., Widmer, H., and Zuber, H. (1978) The complete amino acid sequence of both subunits of C-phycocyanin from the Cyanobacterium *Mastigocladus laminosus*. *Hoppe-Seyler's Z. Physiol. Chem.*, **359**, 1491–1507.

Frank, G. and Weeds, A. G. (1974) The amino acid sequence of the alkali light chains of rabbit skeletal muscle myosin. *Eur. J. Biochem.*, **44**, 317–334.

Geddes, A. J., Graham, G. N., Morris, H. R., Lucas, F., Barber, M., and Wolstenholme, W. A. (1969) Mass-spectrometric determination of the amino acid sequences in peptides isolated from the protein silk fibroin of *Bombyx mori*. *Biochem J.*, **114**, 695–702.

Gray, W. R. and Hartley, B. S. (1963) A fluorescent end-group reagent for proteins and peptides. *Biochem. J.*, **89**, 59P.

Gray, W. R. and Smith, J. F. (1970) Rapid sequence analysis of small peptides. *Anal. Biochem.*, **33**, 36–42.

Gregoire, J. and Rochat, H. (1977) Amino acid sequences of Neurotoxins I and III of the Elapidae snake *Naja mossambica mossambica. Eur. J. Biochem.*, **80**, 283–293.

Gregory, H. (1975) The preparation of deslysylalanyl porcine and bovine insulins. *FEBS Letters*, **51**, 201–205.

Gregory, H. and Preston, B. M. (1977) The primary structure of human urogastrone. *Int. J. Peptide Protein Res.*, **9**, 107–118.

Gross, E. and Witkop, B. (1962) Nonenzymatic cleavage of peptide bonds: the methionine residues in bovine pancreatic ribonuclease. *J. Biol. Chem.*, **237**, 1856–1860.

Hakomori, S. I. (1964) A rapid permethylation of glycolipid and polysaccharide catalyzed by methyl sulfonyl carbanion in dimethyl sulfoxide. *J. Biochem.*, **55**, 205–208.

Hartley, B. S. (1970) Strategy and tactics in protein chemistry. *Biochem J.*, **119**, 805–822.

Hase, T., Matsubara, H., and Evans, M. C. W. (1977) The amino acid sequence of *Chromatium vinosum* ferredoxin: revisions. *J. Biochem.*, **81**, 1745–1749.

Hase, T., Wada, K., Ohmiya, M., and Matsubata, H. (1976) Amino acid sequence of a major component of *Nostoc muscorum* ferredoxin. *J. Biochem.*, **80**, 993–999.

Hase, T., Wakabayashi, S., Matsubara, H., Evans, M. C. W., and Jennings, J. V. (1978) Amino acid sequence of a ferredoxin from *Chlorobium thiosulfatophilum* strain Tassajara, photosynthetic green sulfur bacterium. *J. Biochem.*, **83**, 1321–1325.

Henderson, L. E., Henriksson, D., and Nyman, P. O. (1976) Primary structure of human carbonic anhydrase C. *J. Biol. Chem.*, **251**, 5457–5463.

Herbrink, P. (1976) The polypeptide chain composition of β-crystallin. *Ph.D. Thesis*, University of Nijmegen.

Hitz, H., Schafer, D., and Wittmann-Liebold, B. (1977) Determination of the complete amino acid sequence of protein S6 from the wild-type and a mutant of *Escherichia coli. Eur. J. Biochem.*, **75**, 497–512.

Hoegaerden, M. van and Strosberg, A. D. (1978) Sequence of a rabbit anti-*Micrococcus lysodeikticus* antibody light chain. *Biochemistry*, **17**, 4311–4317.

Hoerman, K. C. and Kamel, K. (1967) Recycled chromatograms for better peptide mapping. *Anal. Biochem.*, **21**, 107–10.

Hogg, R. W. and Hermodson, M. A. (1977) Amino acid sequence of the L-arabinose-binding protein from *Escherichia coli* B/r. *J. Biol. Chem.*, **252**, 5135–5141.

Houmard, J. and Drapeau, G. R. (1972) Staphylococcal protease: a proteolytic enzyme specific for glutamoyl bonds. *Proc. Natn. Acad. Sci. U.S.A.*, **69**, 3506–3509.

Howard, N. L., Joubert, F. J., and Strydom, D. J. (1974) The amino acid sequence of ostrich (*Struthio camelus*) cytochrome c. *Comp. Biochem. Physiol.* **48B**, 75–85.

Hurrell, J. G. R. and Leach, S. L. (1977) The amino acid sequence of soybean leghaemoglobin c_2. *FEBS Letters*, **80**, 23–26.

Inagami, T. (1973) Simultaneous identification of PTH derivatives of histidine and arginine by thin-layer chromatography. *Anal. Biochem.*, **52**, 318–321.

Inglis, A. S. and Burley, R. W. (1977) Determination of the amino acid sequence of Apovitellenin I from duck's egg yolk using an improved sequenator procedure: a comparison with other avian species. *FEBS Letters*, **73**, 33–37.

146

Ingram, V. M. (1953) Phenylthiohyantoins from serine and threonine. *J. Chem. Soc.*, **1953**, 3717–3718.

Isobe, T., Black, L. W., and Tsugita, A. (1977b) Complete amino acid sequence of bacteriophage T_4 internal protein I and its cleavage site on virus maturation. *J. Mol. Biol.*, **110**, 165–177.

Isobe, T., Nakajima, T., and Okuyama, T. (1977a) Reinvestigation of extremely acidic peptides in bovine brain. *Biochim. Biophys. Acta*, **494**, 222–232.

Isobe, T. and Okuyama, T. (1978) The amino acid sequence of S-100 protein (PAP I-b Protein) and its relation to the calcium binding proteins. *Eur. J. Biochem.*, **89**, 379–388.

Jabusch, J. R., Parmelee, D. C., and Deutsch, H. F. (1978) The effect of thiodiglycol and dithiothreitol on the alkaline hydrolysis products of certain amino acid phenylthiohydantoins. *Anal. Biochem.*, **91**, 532–542.

Jackson, R. L., Lin, H.-Y., Chan, L., and Means, A. R. (1977) Amino acid sequence of a major apoprotein from hen plasma very low density lipoprotein. *J. Biol. Chem.*, **252**, 250–253.

Jeppsson, J. and Sjöquist, J. (1967) Thin-layer chromatography of PTH amino acids. *Anal. Biochem.*, **18**, 264–269.

Joassin, L. and Gerday, C. (1977) The amino acid sequence of the major parvalbumin of the whiting (*Gadus merlangus*). *Comp. Biochem. Physiol.* **57B**, 159–161.

Jones, M. G. (1976) Structure and biological activity of peptides. *Ph.D. Thesis*, University of Salford.

Jörnvall, H. (1977) The primary structure of yeast alcohol dehydrogenase. *Eur. J. Biochem.*, **72**, 425–442.

Kasper, C. B. (1975) Fragmentation of proteins for sequence studies and separation of peptide mixtures, in *Protein Sequence Determination*, Ed. S. B. Needleman (Springer-Verlag, Berlin and New York), pp 114–161.

Kawasaki, I. and Itano, H. A. (1972) Methanolysis of the pyrrolidone ring of amino-terminal pyroglutamic acid in model peptides. *Anal. Biochem.*, **48**, 546–556.

Kondo, K., Narita, K., and Lee, C. (1978) Amino acid sequences of the two polypeptide chains in Bungarotoxin from the venom of *Bungarus multicinctus*. *J. Biochem.*, **83**, 101–115.

Kootstra, A. and Bailey, G. S. (1978) Primary structure of histone H2B from Trout (*Salmo trutta*) testes. *Biochemistry*, **17**, 2504–2509.

Krutzsch, H. C. and Kindt, T. J. (1979) The identification of trimethylsilylated dipeptides with chemical ionization mass spectrometry. *Anal. Biochem.*, **92**, 525–531.

Kulbe, K. D. (1974) Micropolyamide thin-layer chromatography of phenylthiohydantoin amino acids (PTH) at subnanomolar level. A rapid micro technique for simultaneous multisample identification after automated Edman degradations. *Anal. Biochem.*, **59**, 564–573.

Laursen, R. A. (1971) Solid-phase Edman degradation: an automatic peptide sequencer. *Eur. J. Biochem.*, **20**, 89–102.

Laursen, R. A. (1972) Automatic solid-phase Edman degradation, in *Methods in Enzymology*, Vol. **XXVB**, 344–359, Eds C. H. W. Hirs and S. N. Timasheff, (Academic Press, London).

Laursen, R. A., Horn, M. J., and Bonner, A. G. (1972) Solid-phase Edman degradation. The use of p-phenyl diisothiocyanate to attach lysine and arginine-containing peptides to insoluble resins. *FEBS Letters*, **21**, 67–71.

Lee, H. and Riordan, J. F. (1978) Does carboxypeptidase Y have intrinsic endopeptidase activity? *Biochem. Biophys. Res. Commun.*, **85**, 1135–1142.

Lin, J. K. and Chang, J. Y. (1975) Chromophoric labeling of amino acids with 4-dimethylaminobenzene-4'-sulfonyl chloride. *Anal. Chem.*, **47**, 1634–1638.

Lindemann, H. and Wittmann-Liebold, B. (1977) Primary structure of protein S13 from the small subunit of *E. coli* ribosomes. *Hoppe-Seyler's Z. Physiol. Chem.*, **358**, 843–863.

Lu, H. and Lo, T. (1978) Complete amino acid sequence of a new type of cardiotoxin of *Bungarus fasciatus* venom. *Int. J. Peptide Protein Res.*, **12**, 181–183.

Macleod, A. R., Wong, N. C. W., and Dixon, G. H. (1977) The amino acid sequence of Trout-testis histone H1. *Eur. J. Biochem.*, **78**, 281–291.

Maeda, N. and Tamiya, N. (1977) Correction of the partial amino acid sequence of Erabutoxins. *Biochem. J.*, **167**, 289–291.

Magnusson, S., Sottrup-Jensen, L., Petersen, T. E., Morris, H. R., and Dell, A. (1974) Primary Structure of the vitamin K-dependent part of prothrombin. *FEBS Letters*, **44**, 189–193.

Mahajan, V. K. and Desiderio, D. M. (1978) Mass spectrometry of acetylated, permethylated and reduced oligopeptides. *Biochem. Biophys. Res. Commun.*, **82**, 1104–1110.

Mak, A. S. and Jones, B. L. (1976) The amino acid sequence of wheat β-purothionin. *Can. J. Biochem.*, **22**, 835–842.

Martinez, G., Kopeyan, C., Schweitz, H., and Lazdunski, M. (1977) Toxin III from *Anemoniasulcata*: primary structure. *FEBS Letters*, **84**, 247–252.

Matsubara, H., Sasaki, R. K., Tsuchiya, D. K., and Evans, M. C. W. (1970) The amino acid sequence of *Chromatium Ferredoxin*. *J. Biol. Chem.*, **245**, 2121–2131.

Matsuo, H., Fujimoto, Y., and Tatsuno, T. (1966) A novel method for the determination of C-terminal amino acids in polypeptides by selective tritium labelling. *Biochem. Biophys. Res. Comm.*, **22**, 69–74.

Meagher, R. B. (1975) Rapid manual sequencing of multiple peptide samples in a nitrogen chamber. *Anal. Biochem.*, **67**, 404–412.

Mendez, E. and Lai, C. Y. (1975) Regeneration of amino acids from thiazolinones formed in the Edman degradation. *Anal. Biochem.*, **68**, 47–53.

Mitchell, W. M. and Harrington, W. F. (1968) Purification and properties of Clostridiopeptidase B (clostripain). *J. Biol. Chem.*, **243**, 4683–4692.

Mok, C. C. and Waley, S. G. (1968) N-terminal groups in lens proteins. *Exp. Eye Res.*, **7**, 148–153.

Morris, H. R. (1972) Studies towards the complete sequence determination of proteins by mass spectrometry: a rapid procedure for the successful permethylation of histidine-containing peptides. *FEBS Letters*, **22**, 257–260.

Morris, H. R. (1975) Protein sequence analysis and the discovery of a new amino acid in prothrombin. *Biochem. Soc. Trans.*, **3**, 465–467.

Morris, H. R., Batley, K. E., and Harding, N. G. L. (1974a) Dihydrofolate reductase: low resolution mass-spectrometric analysis of an elastase digest as a sequencing tool. *Biochem. J.*, **137**, 409–411.

Morris, H. R., Dickinson, R. J. and Williams, D. H. (1973) Studies towards the complete sequence determination of proteins by mass spectrometry; derivatisation of methionine, cysteine and arginine containing peptides. *Biochem. Biophys. Res. Commun.*, **51**, 247–255.

Morris, H. R., Thompson, M. R., Osuga, D. T., Ahmed, A. I., Chan, S. M., Vandenheede, J. R., and Feeney, R. E. (1978) Antifreeze glycoproteins from blood of an antarctic fish. *J. Biol. Chem.*, **253**, 5155–5162.

Morris, H. R., Williams, D. H., and Ambler, R. P. (1971) Determination of the sequence of protein-derived peptides and peptide mixtures by mass spectrometry. *Biochem. J.*, **125**, 189–201.

Morris, H. R., Williams, D. H., Midwinter, G. G., and Hartley, B. S. (1974b) A mass-spectrometric sequence study of the enzyme ribitol dehydrogenase from *Klebsiella aerogenes*. *Biochem. J.*, **141**, 701–713.

Mosesson, M. W., Finlayson, J. S., and Galanakis, D. K. (1973) The essential covalent structure of human fibrinogen evinced by analysis of derivatives formed during plasmic hydrolysis. *J. Biol. Chem.*, **248**, 7913–7929.

Naider, F. and Bohak, Z. (1972) Regeneration of methionyl residues from their sulfonium salts in peptides and proteins. *Biochemistry*, **11**, 3208–3211.

Narita, K. (1958) Isolation of an acetyl peptide from enzymic digests of TMV-protein. *Biochim. Biophys. Acta*, **28**, 184–191.

Nau, H. (1974) New dideutero-perfluoroalkylated oligopeptide derivatives for protein sequencing by gas chromatography–mass spectrometry. *Biochem. Biophys. Res. Commun.* **59**, 1088–1096.

Nau, H. and Bieman, K. (1976) Amino acid sequencing by gas chromatography–mass spectrometry using trifluoro-deuteroalkylated peptide derivatives. *Anal. Biochem.*, **73**, 175–186.

Niall, H. D. (1977) Pehr Edman—Obituary. *Nature*, **268**, 279–280.

Niall, H. D., Jacobs, J. W., van Rietschoten, J., and Tregear G. W. (1974) Protected Edman degradation: a new approach to microsequence analysis of proteins. *FEBS Letters*, **41**, 62–64.

Niketic, V., Thomsen, J., and Kristiansen, K., (1974) Modification of cysteine residues with sodium 2-bromoethane sulphonate. *Eur. J. Biochem.*, **46**, 547–551.

O'Donnell, I. J. and Inglis, A. S. (1974) Amino acid sequence of a feather keratin from Silver Gull (*Larus novae-hollandiae*) and comparison with one from Emu (*Dromaius novae-hollandiae*). *Aust. J. Biol. Sci.*, **27**, 369–382.

Orden, H. O. van and Carpenter, F. H. (1964) Hydrolysis of phenylthiohydantoins of amino acids. *Biochem. Biophys. Res. Commun.*, **14**, 399–403.

Orlov, V. M., Varshavsky, Y. M., and Kiryushkin, A. A. (1972) Comparative studies on photo-ionization and electron-impact ionization of peptide derivatives. *Organic Mass Spectrometry*, **6**, 9–20.

Ovchinnikov, Yu. A., Egorov, C. A., Aldanova, N. A., Feigina, M. Yu., Lipkin, V. M., Abdulaev, N. G., Grishin, E. V., Kiselev, A. P., Modyanov, N. N., Braunstein, A. E., Polyanovsky, O. L., and Nosikov, V. V. (1973) The complete amino acid sequence of cytoplasmic aspartate aminotransferase from pig heart. *FEBS Letters*, **29**, 31–34.

Ovchinnikov, Yu. A., Lipkin, V. M., Modyanov, N. N., Chertov, O. Yu., and Smirnov, Yu. V. (1977) Primary structure of α-subunit of DNA-dependent RNA polymerase from *Escherichia coli*. *FEBS Letters*, **76**, 108–111.

Parham, M. E. and Loudon, G. M. (1978) A new method of determination of the carboxyl-terminal residue of peptides. *Biochem. Biophys Res. Commun.*, **80**, 7–13.

Partridge, S. M. and Davis, H. F. (1950) Preferential release of aspartic acid during the hydrolysis of proteins. *Nature*, **165**, 62–63.

Patthy, L. and Smith, E. L. (1975) Identification of functional arginine residues in ribonuclease A and lysozyme. *J. Biol. Chem.*, **250**, 565–569.

Percy, M. E. and Buchwald, B. M. (1972) A manual method sequential Edman degradation followed by dansylation for the determination of protein sequences. *Anal. Biochem.*, **45**, 60–67.

Peterson, J. D., Nehrlich, S., Oyer, P. E., and Steiner, D. F. (1972) Determination of the amino acid sequence of the monkey, sheep and dog proinsulin C-peptides by a semi-micro Edman degradation procedure. *J. Biol. Chem.*, **247**, 4866–4871.

Pettigrew, G. W. and Smith, G. M. (1977) Novel N-terminal protein blocking group identified as dimethylproline. *Nature*, **265**, 661–662.

Pisano, J. J., Bronzert, T. J., and Brewer, H. B. (1972) Advances in the gas chromatographic analysis of amino acid phenyl- and methylthiohydantoins. *Anal. Biochem.*, **45**, 43–59.

Pisano, J. J., van den Heuvel, W. J. A., and Horning, E. C. (1962) Gas chromatography of phenylthiohydrantoin and dinitrophenyl derivatives of amino acids. *Biochem. Biophys. Res. Commun.*, **7**, 82–86.

Podell, D. N. and Abraham, G. N. (1978) A technique for the removal of pyroglutamic acid from the amino terminus of proteins using calf liver pyroglutamate amino peptidase. *Biochem. Biophys. Res. Commun.*, **81**, 176–185.

Polan, M. L., McMurray, W. J., Lipsky, S. R., and Lande, S. (1970) Mass spectrometry of cysteine-containing peptides. *Biochem. Biophys. Res. Commun.*, **38**, 1127–1133.

Ponstingl, H., Nieto, A., and Beato, M. (1978) Amino acid sequence of progesterone-induced rabbit uteroglobin. *Biochemistry*, **17**, 3908–3912.

Previero, A., Coletti-Previero, M. A., and Axelrud-Cavadore, C. (1967) Prevention of cleavage next to tryptophan residues during the oxidative splitting by N-bromosuccinimide of tyrosyl peptide bonds in proteins. *Arch. Biochem. Biophys.*, **122**, 434–438.

Previero, A., Derancourt, J., Coletti-Previero, M. A., and Laursen, R. A. (1973) Solid phase sequential analysis: specific linking of acidic peptides by their carboxyl ends to insoluble resins. *FEBS Letters*, **33**, 135–138.

Priddle, J. D. and Offord, R. E. (1974) The active centre of triose phosphate isomerase from chicken breast muscle. *FEBS Letters*, **39**, 349–352.

Prox, A. and Weygand, F. (1967) Sequenzanalyse von Peptiden durch Kombination von Gaschromatographie und Massenspektrometrie, in *Peptides* ed. H. C. Beyerman, A. van de Linde, and W. M. van den Brink (North-Holland Publishing Co., Amsterdam) pp 158–172.

Rees, M. W., Short, M. N., Self, R., and Eagles, J. (1974) The amino acid sequences of the tryptic peptides of the cowpea strain of tobacco mosaic virus protein. *Biomed. Mass Spectrom.*, **1**, 237–251.

Richardson, C., Behn'.e, W. D., Freisheim, J. H., and Blumenthal, K. M. (1978) The complete amino acid sequence of the α-subunit of pea lectin. *Biochim. Biophys. Acta*, **537**, 310–319.

Rinderknecht, E. and Humbel, R. E. (1978) The amino acid sequence of human insulin-like growth factor I and its structural homology with proinsulin. *J. Biol. Chem.*, **253**, 2769–2776.

Rochat, H., Bechis, G., Kopeyan, C., Gregoire, J., and van Rietschoten, J. (1976) Use of parvalbumin as a protecting protein in the sequenator: An easy and efficient way for sequencing small amounts of peptides. *FEBS Letters*, **64**, 404–408.

Rodkey, J. A. and Bennett, C. D. (1976) Micro Edman degradation: the use of high-pressure chromatography and gas chromatography in the amino terminal sequence determination of 8 nanomoles of dihydrofolate reductase from a mouse sarcoma. *Biochem. Biophys. Res. Commun.*, **72**, 1407–1413.

Roepstorff, P., Norris, K., Severinsen, S., and Brunfeldt, K. (1970) Mass spectrometry of peptide derivatives. Temporary protection of methionine as sulfoxide during permethylation. *FEBS Letters*, **9**, 235–238.

Roseau, G. and Pantel, P. (1969) Revelation coloree des spots de phenylthio hydantoin d'acides amines. *J. Chrom.*, **44**, 392–395.

Rosmus, J. and Deyl, Z. (1972) Chromatography of N-terminal amino acids and derivatives. *J. Chrom.*, **70**, 221–339.

Sanger, F. (1945) The free amino groups of insulin. *Biochem. J.*, **39**, 507–515.

Sargent, J. R. (1965) *Methods in Zone Electrophoresis* (B. D. H. Chemicals Ltd, Poole, Dorset).

Scheffer, A. J. and Beintema, J. J. (1974) Horse pancreatic ribonuclease. *Eur. J. Biochem.*, **46**, 221–233.

Schroeder, W. A. (1972) Separation of peptides by chromatography on columns of Dowex 50 with volatile developers, in *Methods in Enzymology*, Vol. **XXVB**, 203–213, Eds C. H. W. Hirs and S. N. Timasheff (Academic Press, London).

Schultz, J., Allison, H., and Grice, M. (1962) Specificity of the cleavage of proteins by dilute acid. *Biochemistry*, **1**, 694–698.

150

Sepulveda, P., Marciniszyn, J., Liu, D., and Tang, J. (1975) Primary structure of porcine pepsin III. *J. Biol. Chem.*, **250**, 5082–5088.

Shechter, Y., Patchornik, A., and Burstein, Y. (1976) Selective chemical cleavage of tryptophanyl peptide bonds by oxidative chlorination with N-chlorosuccinimide. *Biochemistry*, **15**, 5071–5075.

Shemyakin, M. M., Ovchinnikov, Yu. A., Kiryushkin, A. A., Vinogradova, E. I., Miroshnikov, A. I., Alakhov, Yu. B., Lipkin, V. M., Shvetsov, Yu. B., Wulfson, N. S., Rosinov, B. V., Bochkarev, V. N., and Burikov, V. M. (1966) Mass spectrometric determination of the amino acid sequence of peptides. *Nature*, **211**, 361–366.

Silver, J. and Hood, L. E. (1974) Automated microsequence analysis in the presence of a synthetic carrier. *Analyt. Biochem.*, **60**, 285–292.

Simon-Becam. A., Claisse, M., and Lederer, F. (1978) Cytochrome c from *Schizosaccharomyces pombe*. *Eur. J. Biochem.*, **86**, 407–416.

Slingsby, C. (1974) Structural studies on γ-Crystallin fraction IV. *D. Phil. Thesis*, University of Oxford.

Slingsby, C. and Croft, L. R. (1978) Structural studies on calf lens γ-Crystallin fraction IV: A comparison of the cysteine-containing tryptic peptides with the corresponding amino acid sequence of γ-Crystallin fraction II. *Exp. Eye Res.*, **26**, 291–304.

Smith, G. P. S. (1978) Sequence of the full length immunoglobulin κ-chain of mouse myeloma MPC11. *Biochem. J.*, **171**, 337–347.

Smithies, O., Gibson, D., Fanning, E. M., Goodfliesh, R. M., Gilman, J. G., and Ballantyne, D. L. (1971) Quantitative procedures for use with the Edman–Begg sequenator. Partial sequences of two unusual immunoglobulin light chains, Rzf and Sac. *Biochemistry*, **10**, 4912–4921.

Stenflo, J., Fernlund, P., Egan, W., and Roepstorff, P. (1974) Vitamin K-dependent modifications of glutamic acid residues in prothrombin. *Proc. Natn. Acad. Sci. U.S.A.* **71**, 2730–2733.

Stenhagen, E. (1961) Massenspektrometrie als Hilfsmittel bei der Strukturbestimmung organischer Verbindungen, besonders bei Lipiden und Peptiden. *Z. Analyt. Chem.*, **181**, 462–480.

Stone, D. and Phillips, A. W. (1977) The amino acid sequence of dehydrofolate reductase from L1210 cells. *FEBS Letters*, **74**, 85–87.

Stone, D., Phillips, A. W., and Burchall, J. J. (1977) The amino acid sequence of the dihydrofolate reductase of a Trimethoprim-resistant strain of *Escherichia coli*. *Eur. J. Biochem.*, **72**, 613–624.

Strickland, M., Strickland, W. N., Brandt, W. F., and von Holt, C. (1977) The complete amino acid sequence of histone $H_2B_{(1)}$ from sperm of the Sea Urchin *Parechinus angulosus*. *Eur. J. Biochem.*, **77**, 263–275.

Strid, L. (1973) Separation of peptides according to charge by gel filtration in presence of ionized detergents. *FEBS Letters*, **33**, 192–196.

Strydom, D. J. (1977) Snake venom toxins. The amino acid sequence of a short-neurotoxin homologue from *Dendroaspis polylepis polylepis* (Black Mamba) venom. *Eur. J. Biochem.*, **76**, 99–106.

Summers, M. R., Smythers, G. W., and Oroszlan, S. (1973) Thin-layer chromatography of sub-nanomole amounts of phenylthiohydantoin (PTH) amino acids on polyamide sheets. *Anal. Biochem.*, **53**, 624–628.

Sutton, M. R., Fall, R. R., Nervi, A. M., Alberts, A. W., Vagelos, P. R., and Bradshaw, R. A. (1977). Amino acid sequence of *Escherichia coli* biotin carboxyl carrier protein (9100). *J. Biol. Chem.*, **252**, 3934–3940.

Tanaka, M., Haniu, M., Yasunobu, K. T., Evans, M. C. W., and Rao, K. K. (1975) The amino acid sequenie of ferredoxin II from *Chlorobium limicola*, a photosynthetic green bacterium. *Biochemistry*, **14**, 1938–1943.

Tanaka, M., Haniu, M., Yasunobu, K. T., and Norton, T. R. (1977) Amino acid sequence of the *Anthopleura xanthogrammica* heart stimulant Anthopleurin A. *Biochemistry*, **16**, 204–208.

Tanaka, M., Haniu, M., Yasunobu, K. T., and Yoch, D. C. (1977) Complete amino acid sequence of Azotoflavin, a flavodoxin from *Azotobacter vinelandii. Biochemistry*, **16**, 3525–3537.

Tarr, G. E. (1975) A general procedure for the manual sequencing of small quantities of peptides. *Anal. Biochem.*, **63**, 361–370.

Tarr, G. E., Beecher, J. F., Bell, M., and McKean, D. J. (1978) Polyquarternary amines prevent peptide loss from sequenators. *Anal. Biochem.*, **84**, 622–627.

Thomas, D. W. (1968) Mass spectrometry of permethylated peptide derivatives: extension of the technique to peptides containing aspartic acid, glutamic acid or tryptophane. *Biochem. Biophys. Res. Commun.*, **33**, 483–486.

Thomas, D. W. (1969) Mass spectrometry of N-permethylated peptide derivatives: artifacts produced by C-methylation. *FEBS Letters*, **5**, 53–56.

Thomas, D. W., Das, B. C., Gero, S. D., and Lederer, E. (1968) Mass spectrometry of permethylated peptide derivatives: extension of the technique to peptides containing arginine or methionine. *Biochem. Biophys. Res. Commun.*, **32**, 519–525.

Thomsen, J., Bucher, D., Brunfeldt, K., Nexp, E., and Olsen, H. (1976) An improved procedure for automated Edman degradation used for determination of the N-terminal amino acid sequence of human transcobalamin I and human intrinsic factor. *Eur. J. Biochem.*, **69**, 87–96.

Tomita, M., Furthmayr, H., and Marchesi, V. T. (1978) Primary structure of human erythrocyte glycophorin A. Isolation and characterization of peptides and complete amino acid sequence. *Biochemistry*, **17**, 4756–4770.

Tschesche, H., Schneider, M., and Wachter, E. (1972) Mass spectral identification and quantification of phenylthiohydantoin derivatives from Edman degradation of proteins. *FEBS Letters*, **23**, 367–370.

Vandekerckhove, J. and van Montagu, M. (1974) Sequence analysis of fluorescamine-stained peptides and proteins purified on a nanomole scale. *Eur. J. Biochem.*, **44**, 279–288.

Vandekerckhove, J. S. and van Montagu, M. C. (1977) sequence of the A-protein of coliphage MS2. *J. Biol. Chem.*, **252**, 7773–7782.

Vetter-Diechtl, H., Vetter, W., Richter, W., and Bieman, K. (1968) Ein fur Massenspektrometrie und Gaschromatographie geeignetes Argininderivat. *Experientia*, **24**, 340–341.

Vilkas, E. and Lederer, E. (1968) N-methylation des peptides par la methode de Hakomori: structure du mycoside C_{b1}. *Tetrahedron Letters*, **1968**, 3089–3092.

Waley, S. G. and Watson, J. (1953) The action of trypsin on polylysine. *Biochem. J.*, **55**, 328–337.

Walker, J. M., Hastings, J. R. B., and Johns, E. W. (1977) The primary structure of a non-histone chromosomal protein. *Eur. J. Biochem.*, **76**, 461–468.

Wang, C., Avila, R., Jones, B. N., and Gurd, F. R. N. (1977) Complete primary structure of the major component myoglobin of Pacific Common Dolphin (*Delphinus delphis*). *Biochemistry*, **16**, 4978–4981.

Weiner, A. M., Platt, T., and Weber, K. (1972) Amino-terminal sequence analysis of proteins purified on a nanamole scale by gel electrophoresis. *J. Biol. Chem.*, **247**, 3242–3251.

Wilkinson, J. M. and Grand, R. J. A. (1978) The amino acid sequence of chicken fast-skeletal-muscle Troponin I. *Eur. J. Biochem.*, **82**, 493–501.

Wilkinson, J. M., Press, E. M., and Porter, R. R. (1966). The N-terminal sequence of the heavy chain of rabbit immunoglobulin IgG. *Biochemical J.*, **100**, 303–308.

152

Wiman, B. (1977) Primary structure of the B-chain of human plasmin. *Eur. J. Biochem.*, **76**, 129–137.

Winkler, H. D. and Beckley, H. D. (1972) Field desorption mass spectrometry of peptides. *Biochem. Biophys. Res. Commun.*, **46**, 391–398.

Wittmann-Liebold, B. (1973) Amino acid sequence studies on ten ribosomal proteins of *Escherichia coli* with an improved sequenator equipped with an automatic conversion device. *Hoppe-Seyler's Z. Physiol. Chem.*, **354**, 1415–1431.

Wittmann-Liebold, B. and Marzinzig, E. (1977) Primary structure of protein L28 from the large subunit of *Escherichia coli*. *FEBS Letters* **81**, 214–217.

Woods, K. R. and Wang, K. T. (1967) Separation of dansyl-amino acids on polyamide layer chromatography. *Biochim. Biophys. Acta*, **133**, 369–370.

Wunderer, G. and Eulitz, M. (1978) Amino acid sequence of toxin I from *Anemonia sulcata*. *Eur. J. Biochem.*, **89**, 11–17.

Zuber, H. (1964) Purification and properties of a new carboxypeptidase from citrus fruit. *Nature*, **201**, 613.

General Index